TERTIARY LEVEL BIOLOGY

Plant Molecular Biology

DONALD GRIERSON, B.Sc., Ph.D.

Reader in Plant Physiology
School of Agriculture
University of Nottingham

SIMON N. COVEY, B.Sc., Ph.D.

Senior Scientific Officer
John Innes Institute
Norwich

Blackie

Glasgow and London
Distributed in the USA by
Chapman and Hall
New York

Blackie & Son Limited
Bishopbriggs, Glasgow G64 2NZ

Furnival House, 14–18 High Holborn, London WC1V 6BX

Distributed in the USA by
Chapman and Hall
in association with Methuen, Inc.
733 Third Avenue, New York, N.Y. 10017

© 1984 Blackie & Son Ltd
First published 1984

British Library Cataloguing in Publication Data

Grierson, D.
 Plant molecular biology.—(Tertiary level
 biology)
 1. Plant physiology   2. Molecular biology
 I. Title   II. Covey, Simon N.   III. Series
 581.8′8   QK711.2

 ISBN 0–216–91631–2
 ISBN 0–216–91632–1 Pbk

For the USA, International Standard Book Numbers are
 0–412–00651–0
 0–412–00661–8 (pbk)

Filmset by Thomson Press (India) Limited, New Delhi
and printed in Great Britain by
McCorquodale (Scotland) Ltd., Glasgow.

Preface

The development of molecular biology has had a tremendous impact on the biological sciences over the last 30 years. Not only has it proved intellectually exciting and provided a fascinating insight into how cells work, but recent advances have been so successful that they are already forming the basis of industrial processes. Most of the early progress was made by research workers who concentrated on animals and micro-organisms, but plant molecular biology has now developed sufficiently for it to be treated as a discipline in its own right.

Plant cells contain DNA in nuclei, plastids and mitochondria, and so offer the unique challenge of studying the interaction of three separate genetic systems in a single organism. Furthermore, since plant development involves co-ordinated gene expression in response to internal and external signals, plant molecular biology can provide a fundamental insight into how development is regulated. It is also of value in breeding programmes, in understanding the interactions between plants and micro-organisms, and may suggest new ways of manipulating plant growth, development and productivity.

Despite the fact that there is considerable interest in the topic, reports of much of the work that has taken place on plants over the last few years have been confined, largely, to scientific journals and highly-priced conference proceedings. This has tended to maintain plant molecular biology as a specialist field and has hindered the wider appreciation of the subject. This volume arose from the conviction that there is a need for a short book that highlights important recent developments in this very exciting area, yet contains the background information useful to students and non-specialists. We have dealt with topics that we feel should form the 'core' of any course on plant molecular biology. We have assumed a basic knowledge of molecular genetics, biochemistry and plant and cell biology, but have included an outline of some important methods used in the study of plant nucleic acids, together with a discussion of the prospects for plant genetic engineering. We believe that this rather limited interpretation of the scope of plant molecular biology is justified on the grounds that it is intellectually the most exciting and it is in these areas where most progress is being made. We have kept references in the text to a minimum, citing recent important papers and review articles, but access to the extensive and rapidly developing literature on the subject is provided in the detailed bibliography at the end of the book.

We are indebted to many colleagues for their invaluable advice and assistance during the preparation of the manuscript. In particular we thank Dr J. A. Downie, Dr J. Burgess, Dr N. J. Brewin, Dr C. J. Leaver, Dr J. L. Firmin, Mr T. Collins, Mr P. G. Scott and Ms K. Stallard.

Donald Grierson
Simon Covey

Contents

Chapter 9   PLANT VIRUSES

Chapter 10   PROSPECTS FOR THE GENETIC ENGINEERING OF PLANTS

REFERENCES

INDEX

CHAPTER ONE

GENE CLONING, IDENTIFICATION AND SEQUENCING

Recombinant DNA technology utilizes a few simple but powerful methods which allow researchers to identify, purify and determine the structure and regulation of genes and their products. It also provides the means for transferring genes from one organism to another and, perhaps, for designing new genes. Unfortunately, this area of research has its own jargon and abbreviations which make it difficult for the uninitiated to make head or tail of what is going on. This chapter has been written with this in mind and a general outline of the procedures is given in order to help the reader interpret the experiments discussed in the rest of the book. Detailed accounts of experimental methods may be found in the excellent reviews of the subject cited at the end of the book.

1.1 Restriction enzymes

Restriction enzymes are essential for the characterization of DNA molecules and for gene cloning and sequencing because they generate small defined fragments of DNA which are easy to manipulate and study. Over 200 restriction endonucleases have been purified from bacteria. Three classes of restriction enzymes are recognized but only Type II enzymes are extensively used for genetic manipulation. The names of the enzymes are derived from the names of the bacteria from which they are purified. For example, *Eco* R1 comes from *Escherichia coli* and *Hae* III comes from *Haemophilus aegyptus*. In nature these enzymes are probably involved in the recognition and destruction of foreign DNA sequences which may enter bacterial cells. They recognize and cleave double-stranded DNA molecules at or near specific base sequences.

Restriction enzyme recognition sites generally consist of 4–6 base pairs with twofold symmetry. Cleavage can occur in the recognition sequence

Table 1.1   Restriction enzyme recognition and cleavage sites

Restriction enzyme	Recognition site	Ends created	
Eco R1	5′ GAATTC 3′ 3′ CTTAAG 5′	G CTTAA 5′	5′ AATTC G
Sma 1	5′ CCCGGG 3′ 3′ GGGCCC 5′	CCC GGG 5′	5′ GGG CCC
Hind 111	5′ AAGCTT 3′ 3′ TTCGAA 5′	A TTCGAA 5′	5′ AGCTT A
Hind 11	5′ GTPyPuAC 3′ 3′ CAPuPyTG 5′	GTPy CAPu 5′	5′ PuAC PyTG
Pst 1	5′ CTGCAG 3′ 3′ GACGTC 5′	CTGCA G 5′	5′ G ACGTC
Hinf 1	5′ GANTC 3′ 3′ CTNAG 5′	G CTNA 5′	5′ ANTC G
Sta N1	5′ NNNNNNNNNNGATGC 3′ 3′ NNNNNNNNNCTACG 5′	N NNNNN 5′	5′ NNNNNNNNNGATGC NNNNNCTACG

itself or some distance away (Table 1.1). The DNA molecule can either be cut symmetrically to produce 'blunt ends', as with *Sma* 1, or asymmetrically to generate staggered cuts with 3′ or 5′ projections. Restriction enzymes such as *Pst* 1 and *Eco* R1, which make staggered cuts within the recognition sequence, generate cohesive or 'sticky' ends, which can base-pair with each other.

The frequency and distribution of restriction enzyme sites within a DNA molecule vary greatly with the sequence being studied. For example, in the repeat unit of the 25S and 18S rRNA genes in wheat (about 9000 base pairs long) there is only one *Eco* R1 site, whereas in the spacer region between the 25S and 18S genes (see Chapter 2) there are at least 17 *Hha* 1 sites. On the other hand, cauliflower mosaic virus DNA (the Strasbourg isolate, 8024 base pairs, see Chapter 9) has only one *Hha* 1 site but seven *Eco* R1 recognition sequences.

Restriction enzymes are generally unable to cut at their recognition sites if specific cytosine (C) or adenine (A) residues are methylated (Table 1.2). This can sometimes complicate the job of restriction enzyme mapping but it

Table 1.2 Effect of methylation on restriction enzyme site recognitions

Restriction enzyme	Recognition sequence	Methylated sequence cleaved	Methylated sequence not recognized
Eco R1	$\overset{\downarrow}{G}$AATTC	—	GAA*TTC
Eco R11	$\downarrow$CC$\overset{A}{\underset{T}{}}$GG	—	CC*$\overset{A}{\underset{T}{}}$GG
Sau 3A1	$\downarrow$GATC	GA*TC	GATC*
Taq 1	$\overset{\downarrow}{T}$CGA	TC*GA	TCGA*

can be advantageous when studying gene regulation. There is evidence that specific methylation of C residues may be involved in rendering genes transcriptionally inactive (see Chapter 3) and restriction enzymes provide a very effective means of detecting changes in methylation patterns related to gene expression.

1.2 Analysis of restriction enzyme digests

Restriction endonuclease fragments are generally fractionated by electrophoresis in agarose gels, to separate the molecules on the basis of molecular weight. The lengths of the fragments are measured by comparing their mobility with that of marker DNAs of defined length included in the gel. Only microgram quantities of DNA are needed for the analysis. The fragments can be visualized in the gel by staining with ethidium bromide and observing the orange fluorescence of the dye under ultraviolet light. If a complex genome is cut with restriction enzymes, many fragments of different lengths are generated and these produce a smear through the gel. However, repeated sequences are often present in the genome and these give rise to many copies of a particular fragment which can be seen as a distinct band. In contrast, restriction enzyme digestion of simple genomes, such as those from chloroplasts or plasmids, generally produces a relatively small number of unique fragments.

Digestion of DNA regions containing blocks of repeating sequences can sometimes lead to the production of 'ladders' when the digest is frac-

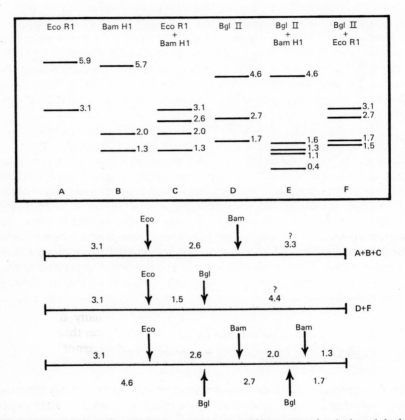

Figure 1.1 Mapping of restriction enzyme sites in a DNA segment by single and double digestions. The DNA molecule in this example is cleaved with *Eco* R1, *Bam* H1 and *Bgl* 11 in separate reactions. Further samples of the DNA are then digested with mixtures of two restriction enzymes. The fragments are separated by electrophoresis in agarose gels and visualized by their fluorescence under ultraviolet light in the presence of ethidium bromide. The molecular weights of the individual fragments, shown $\times 10^{-6}$, are calculated from the mobility of known marker DNAs fractionated in the same gel. Partial maps can easily be made from the data. For example, the relative positions of the *Eco* R1 site and one of the two *Bam* H1 sites can be deduced from the information in A, B and C. Similarly, one of the *Bgl* 11 sites can be positioned relative to the *Eco* R1 site from the information in D and F. The problem is to locate the second *Bgl* 11 and *Bam* H1 sites to the right of the diagram. The solution is found by ordering the overlapping fragments shown in B, D and E. The order is confirmed by the information in F.

tionated by gel electrophoresis. There are two ways in which this can come about. Incomplete digestion of a block of repeating sequences can lead to the production of multimers of the repeating unit; in this situation, incubation with more restriction enzyme or for a longer time results in complete digestion to the basic repeating sequence. Frequently, however, the repeating sequences are not perfect and some restriction sites are altered, by base changes, so that the enzyme no longer recognizes or cuts the DNA at a particular site. This also produces ladders but the patterns are unaffected by prolonged digestion.

A physical map of a particular DNA region can be obtained by locating internal restriction enzyme sites and determining the order of these, as illustrated in Figure 1.1. Restriction enzyme maps of large DNA molecules are constructed by first cutting them into smaller fragments and working out the order of internal restriction sites in each fragment. Digestion of the DNA with different enzymes produces a new series of overlapping fragments which allows the complete map to be assembled. Once the physical location of specific fragments is determined, they can be cut out of the DNA molecule and their properties and function investigated. For example, DNA–RNA hybridization can be used to locate regions coding for rRNA or mRNA (Chapter 2), and it can be used in coupled transcription–translation systems to locate protein-coding regions (Chapter 4) or for cloning or DNA sequencing.

1.3 Mapping of specific nucleotide sequences to individual restriction enzyme fragments

After electrophoresis, DNA fragments can be transferred directly from the gel and fixed to a sheet of nitrocellulose paper, DBM-paper (diazobenzyloxymethyl-paper), or some other suitable support, preserving the original banding pattern of the DNA. This procedure, known as 'Southern blotting' (named after the man who invented it) involves a high-pH denaturation step so that the DNA is attached to the paper in a single-stranded state. This allows the immobilized DNA to form hybrids when the paper is soaked in solutions of radioactive complementary nucleic acid probes. Large numbers of DNA fragments can be transferred to a sheet of paper and challenged in this way to locate regions coding for specific nucleotide sequences. An analogous procedure, used to transfer RNA molecules from gels to nitrocellulose or DBM-paper, known as 'Northern blotting' enables the detection and study of RNA sequences which have homology with specific DNA probes. DNA probes are generally made radioactive *in vitro* by 'nick translation'. In this procedure unlabelled double-stranded DNA of interest is nicked with deoxyribonuclease I, to introduce single-stranded breaks in the polynucleotide chain. DNA polymerase I is then used to incorporate ^{32}P-or ^{35}S-labelled deoxyribonucleotides into the DNA at the site of the nicks. DNA probes of extremely high specific radioactivity can be generated by this means since, starting from one nick, the DNA polymerase enzyme catalyses the introduction of many radioactive nucleotides. RNA molecules are also used as probes and can be labelled *in vitro* by several methods, including the addition of radioactive ribonucleotides to the 5′ end with the enzyme polynucleotide kinase.

1.4 Gene cloning

Gene cloning provides a means of purifying and propagating specific DNA segments. Large DNA molecules are first dissected with restriction enzymes to produce specific fragments. These are then inserted into a cloning vector, which is capable of being replicated in *E. coli* or some other suitable host, by recombination *in vitro*. The chimaeric molecule, containing the vector DNA and inserted foreign DNA, is introduced into bacterial cells where it multiplies. Many copies of the inserted DNA can subsequently be recovered by purifying the hybrid vector from the cells. Cloning is generally carried out with a complex mixture of foreign DNA sequences. Thousands of recombinant DNA molecules are generated and these are replicated individually in bacterial cells to produce a 'library' of different DNA clones. The cloned DNA is not necessarily expressed in the bacteria. Special screening methods have to be devised in order to identify clones of particular interest. Cloning can be carried out in other bacteria and yeast but the systems based on *E. coli* are the most highly developed. Broadly speaking there are three types of cloning vectors for use in *E. coli* cells: plasmids, bacteriophage lambda and cosmids, discussed below.

Plasmids

Plasmids are extra-chromosomal, double-stranded, circular DNA molecules found in prokaryotic and eukaryotic cells. They carry genes for DNA replication and segregation and also frequently contain other genes for heavy metal tolerance, antibiotic resistance or the ability to metabolize exotic organic compounds. In *Rhizobium*, which forms root nodules in association with leguminous plants (Chapter 7), the genes for nitrogen

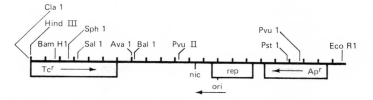

Figure 1.2 Physical map of plasmid pBR322 showing restriction enzyme sites. The map is actually circular but is shown in a linear form with the *Hind* III site at the start of the Tc^r (tetracycline resistance) gene on the left. Ap^r is the ampicillin resistance gene; *ori* is the origin of replication. The direction of transcription of particular genes is shown by the arrows. Modified from Thompson (1982).

fixation are carried on a plasmid. In *Agrobacterium*, which causes plant cancers, the tumour genes are carried on a plasmid and can be transferred to the plant genome and stably inherited (Chapter 8). Plasmids are also found in plant cell organelles (Chapter 5). Some plasmids are extremely large but small plasmids are more suitable for genetic manipulation experiments and naturally-occurring plasmids have been modified for this purpose. They are designated by 'p' for plasmid, followed by the initials of the research worker, laboratory or organism they came from, plus a strain number. One engineered plasmid which is very frequently used for cloning in *E. coli* is pBR322 (Figure 1.2). The entire plasmid DNA consists of 4362 base pairs and the complete sequence is known. It carries genes for resistance to the antibiotics tetracycline and ampicillin and has a number of single restriction enzyme sites into which foreign DNA can be inserted. Some of the restriction sites are in the antibiotic-resistance genes (Figure 1.2) and cloning into these sites inactivates these genes, a phenomenon which is useful during selection for recombinants. Plasmids are introduced into *E. coli* cells by transformation. The bacterial cells are rendered 'competent' by washing them in ice-cold MgCl$_2$ and incubating them in CaCl$_2$ overnight. Subsequently, approximately 20% of the surviving cells take up added DNA. Transformation efficiency is greatest with circular plasmids containing only small inserts of passenger DNA. For this reason vectors other than plasmids are often favoured for cloning long stretches of DNA.

Bacteriophage lambda

The lambda genome consists of a double-stranded linear DNA molecule of approximately 50 kb (50 000 base pairs), which is packaged into the bacteriophage particle. The 5′ ends of the genome have 12-base projections which are complementary and form 'sticky ends'. This allows the bacteriophage DNA to form a circle after infection of *E. coli*. Only about half of the genes are essential for bacteriophage growth and plaque formation and consequently the remaining genes can be removed and replaced with foreign DNA. A number of restriction enzyme sites are available for doing this. The recombinant DNA molecules can be mixed with cell extracts containing bacteriophage proteins and packaged with high efficiency into phage particles *in vitro*, providing the total length of the DNA is between 78–105% of the wild type. DNA molecules outside this size range will not package and this automatically discriminates against non-recombinant DNAs and selects for recombinant molecules. Because of the limitations imposed by the packaging requirement, lambda vectors cannot be used to clone more than 18–21 000 base pairs of foreign DNA. Introduction of the recombinant DNA into *E. coli* cells (transduction) occurs by the normal bacteriophage infection mechanism and is very efficient.

Cosmids

These are hybrid cloning vehicles which combine some of the advantages of plasmids and lambda phage. Cosmid vectors contain the cohesive (*cos*) sites, or 'sticky ends' of lambda and plasmid DNA. They can be packaged into lambda bacteriophage particles *in vitro*. However, most of the lambda DNA is removed, allowing for up to 52 000 base pairs of foreign DNA to be inserted. The packaged DNA can be introduced very efficiently into *E. coli* cells by transduction. Once inside the cells the cosmid replicates as a plasmid.

Recombination in vitro

An outline of the procedure for introducing a piece of foreign DNA into a cloning vector is shown in Figure 1.3. The vector and the DNA to be cloned are cut with the same restriction enzyme to produce sticky ends. If a

Figure 1.3 Insertion of a DNA restriction enzyme fragment into a cloning vector. The vector and DNA containing the gene to be cloned are digested with the same restriction enzyme, producing identical, overlapping, sticky ends. The two DNA preparations are annealed so that the sticky ends can base-pair with one another. (Note that the association of the vector and the DNA fragment to be cloned is only one of several possible combinations). The hybrid molecules are then covalently linked (ligated) by adding T4 DNA ligase and ATP. The region of the plasmid DNA marked 'Res' represents an antibiotic-resistance gene.

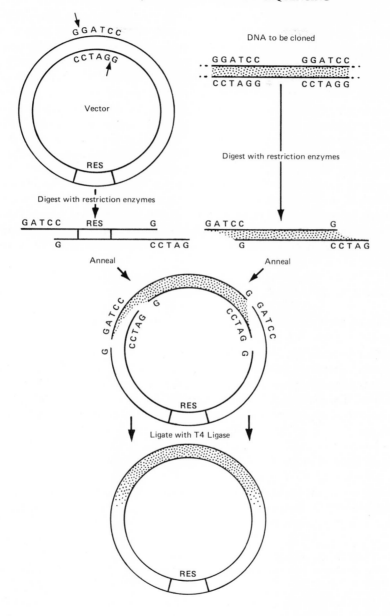

particular restriction enzyme cuts in the middle of a gene, thereby inactivating it, a different restriction enzyme is used. The passenger DNA and the vector are mixed and complementary ends allowed to base pair. The ends are then covalently joined by T4 DNA ligase in the presence of ATP. It is also possible to covalently link blunt-ended DNA molecules with T4 ligase, but this is less efficient than with sticky ends. After recombination *in vitro*, the vector is introduced into *E. coli* cells by transformation (for plasmids) or transduction (for bacteriophage lambda and cosmids).

The efficiency of recombination can be increased in several ways. Self-ligation of the cloning vector can be prevented by removing the 5′ phosphates with bacterial alkaline phosphatase (sometimes called 'bapping') so that they do not act as a substrate for T4 DNA ligase. When this is done only the 5′ phosphate of the passenger DNA can be ligated to the 3′ OH of the vector. This results in nicked circular DNA which can be used successfully for cloning. Alternatively, 3′ homopolymer 'tails' or 'linkers' of G or C oligomers can be added to the vector and passenger DNA, using the enzyme deoxynucleotidyl transferase. If the vector is given G tails and the DNA to be cloned is given C tails then they can base pair only with each other, not with themselves.

cDNA cloning

Plant nuclear DNA is extremely complex (Chapter 2) and a library of genomic clones is large and difficult to screen for individual genes of interest. The cloning of DNA complementary to mRNA sequences (cDNA) generates specific probes which can be used to provide useful information about mRNA sequences and the genes encoding them.

Most plant mRNAs encoded by nuclear DNA and translated in the cytosol (the cytoplasmic compartment, excluding the membrane-bounded organelles such as plastids and mitochondria) have poly-A sequences from 50–200 nucleotides long at their 3′ ends. Not all copies of mRNA molecules coding for a particular protein have these poly-A sequences (Chapter 3), but sufficient do for them to be used as templates for cDNA synthesis *in vitro*. The first step is to extract the mRNA. Sometimes total cellular RNA is used as starting material but in many cases specific mRNA sequences can be enriched for by preparing the RNA from subcellular fractions. For example, the cytosol polyribosomes of leaves are used to obtain the mRNAs transcribed from nuclear genes encoding chloroplast proteins (Chapter 4), whereas the membrane-bound ribosomes of seed endosperm or cotyledons are a source of the mRNAs for seed storage proteins

(Chapter 3). Enrichment for particular mRNAs can also be carried out by fractionating the RNA by size, using centrifugation or electrophoresis. The poly-A$^+$ mRNA is purified by poly(U)-sepharose or oligo(dT)-cellulose affinity chromatography and copied into cDNA *in vitro* with reverse transcriptase from avian myeloblastosis virus. A synthetic sequence of oligo(dT) from 12–18 nucleotides long is added as a primer for the reaction. This base-pairs with the poly-A region of the mRNA and in the presence of deoxynucleotide triphosphates is elongated in the $5' \rightarrow 3'$ direction by the enzyme (Figure 1.4). At the end of the reaction, the RNA strand in the DNA–RNA hybrid is digested with alkali, leaving the single-stranded DNA unaffected. A region near the 3' end of the cDNA frequently forms a base-paired stem with part of the internal sequence of the same strand. This can then be used as a primer for the *E. coli* DNA polymerase I enzyme, which elongates the DNA to form a double-stranded hairpin (Figure 1.4). The loop in the cDNA is then digested with S1 nuclease from *Aspergillus*

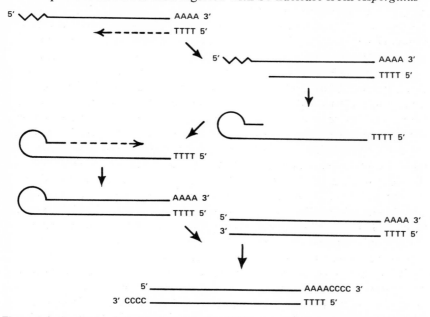

Figure 1.4 Synthesis of double-stranded copy DNA from mRNA. Poly-A-containing mRNA is primed with a short oligo(dT) primer and a cDNA strand synthesized by reverse transcriptase. The original RNA strand is removed by alkali and the single-stranded DNA is then elongated from a naturally-occurring 3' loop which functions as a primer for DNA polymerase. The double-standed DNA is then digested with S1 nuclease to remove the single-stranded loop. G or C tails may be added for recombination *in vitro*.

oryzae, which under appropriate conditions attacks only single-standed DNA. This produces a double-stranded cDNA sequence with free 3' and 5' ends. At this stage the cDNA may be fractionated by sucrose gradient centrifugation to purify the longest copies since the reaction rarely generates a full-length cDNA. Starting with 2–5 micrograms of poly-A$^+$ mRNA, between 20–100 nanograms of double stranded cDNA may be produced by this procedure. Linkers are sometimes added to the 3' ends of the cDNA and complementary linkers attached to the 3' ends of the insertion site in the cloning vector. The vector and cDNA are then annealed, covalently joined with T4 DNA ligase and used for cloning (see *Recombination in vitro*, above).

1.5 Screening of clones

E. coli cells transformed with recombinant plasmids are generally selected using antibiotic-resistance markers. In the case of pBR322 the plasmid carries genes for resistance to ampicillin and tetracycline (Figure 1.2). If cloned DNA is inserted in the tetracycline-resistance gene, for example in the *Bam* H1 or *Sal* 1 sites, the gene is inactivated. Bacteria carrying recombinant pBR322 plasmids are thus resistant to ampicillin but sensitive to tetracycline. They can be selected in one of two ways. In the first method transformed cells are allowed to grow and then plated on agar containing ampicillin. The colonies are then replica-plated on to medium containing tetracycline, either using a velvet pad to transfer the colonies or by moving them manually with a toothpick. Those colonies which grow on ampicillin but not on tetracycline contain the inserted DNA. In the second method, the cells are treated with tetracycline, which prevents those containing recombinants from growing. A second antibiotic, D-cycloserine, is then added, which kills all growing cells. When the cultures are subsequently plated out on ampicillin, all those that grow contain recombinant plasmids.

Bacterial colonies containing cloned DNA can be rapidly screened for specific sequences if a suitable DNA or RNA hybridization probe is available. The colonies are replica-plated onto nitrocellulose filters which are placed on nutrient agar and the colonies allowed to grow. The nitrocellulose filters are then placed on wet filter paper containing alkali, which lyses the cells and denatures the DNA. The single-stranded DNA is then baked on the filters, which are subsequently washed and incubated in a small volume of radioactive probe DNA or RNA. Individual colonies which contain sequences complementary to the radioactive probe form hybrids which are detected by washing the filters, drying them and exposing

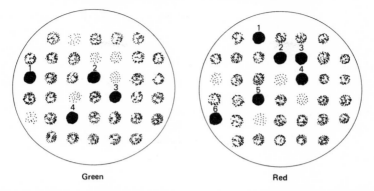

Green   Red

Figure 1.5 Detection of cloned DNA sequences by colony hybridization. Bacterial colonies containing cloned genes are grown on nitrocellulose filters, the colonies lysed and single-stranded DNA immobilized on the filters. The filters are then immersed in a small volume of radioactive nucleic acid probe which is allowed to hybridize to the DNA on the filters. After washing, colonies which contain DNA sequences complementary to the probe are detected by autoradiography. The example illustrates replicated groups of tomato clones hybridized with a radioactive nucleic acid probe complementary to mRNA from either red or green fruit.

to X-ray film (Figure 1.5). This procedure can be used to select colonies which hybridize to cDNA copied from mRNA from one type of cell rather than another (for example light-grown leaves but not dark-grown leaves) or to a purified mRNA fraction, or to a synthetic oligonucleotide probe the sequence of which corresponds to part of the amino acid sequence of a particular protein.

Genes coding for specific proteins can also be detected under favourable circumstances. Some cloning vectors place inserted DNA under the control of bacterial sequences called promoters, which regulate transcription. If the inserted sequence is in the correct reading frame it can be expressed and may direct the synthesis of part of a protein. If a suitable antibody is available, the colonies can be lysed and screened for a specific immunoprecipitation reaction.

Colonies can also be screened for sequences related to specific mRNAs by hybrid-arrest and hybrid-release translation (Figure 1.6). In hybrid-arrest translation, DNA prepared from individual cDNA clones is denatured and allowed to hybridize to mRNA preparations. The mRNA is then added to an *in-vitro* protein synthesis system. The radioactive proteins synthesized are fractionated by sodium dodecylsulphate (SDS)-polyacrylamide gel electrophoresis and detected by autoradiography or a more sensitive variation of this called fluorography. Comparison of the

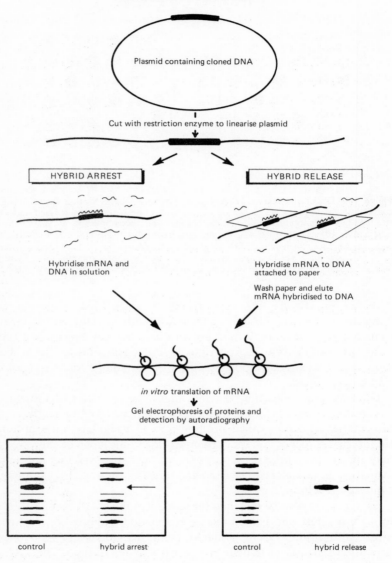

Figure 1.6 An outline of the use of cloned DNA for the identification or purification of mRNA by hybrid-arrest and hybrid-select translation.

mRNA translation products obtained before and after hybridization identifies a clone containing DNA complementary to a specific mRNA because hybridization prevents translation of that mRNA (Figure 1.6). Hybrid-release translation (also called hybrid-select translation) can be used to identify a particular DNA clone and also to purify an individual mRNA. The cloned DNA is denatured and attached to nitrocellulose or DBM paper. The immobilized DNA is then hybridized with total mRNA, washed and the purified mRNA 'melted' off the paper, recovered and translated *in vitro*. The results are again analysed by SDS-gel electrophoresis of the mRNA translation products, followed by autoradiography (Figure 1.6).

In addition to their use in the identification of mRNA molecules and in studying gene expression in bacteria, cDNA clones can also be used for DNA sequencing, which can provide a partial or complete amino acid sequence of the protein for which it codes, or for the identification of genomic clones. The isolation of genomic clones is important since this makes it possible to study the control regions surrounding genes and also to study the sequence organization of the gene itself, which is frequently more complex than that present in the cDNA copies of the corresponding mRNA (Chapter 2).

1.6 DNA sequencing

There are two methods of DNA sequencing in common use: the 'chain terminator' method developed by Sanger and his associates and the chemical method devised by Maxam and Gilbert. Both depend upon the production of a series of radioactive single-standed DNA fragments which start from the same nucleotide but which vary in length. Four sets of fragments are produced for each DNA sequence, each ending with one (or one out of two) of the four bases. The four sets of fragments are then separated into parallel sequence 'ladders' by high resolution electrophoresis on long, ultrathin polyacrylamide gels and detected by autoradiography with X-ray film. Each rung of the ladder represents a fragment which differs in length from its neighbour by one nucleotide. The base which terminates each fragment is identified by the reaction conditions used to generate each of the four gel tracks and the DNA sequence can be read from the autoradiograph (Figure 1.7).

In the Maxam and Gilbert method, restriction enzyme fragments of DNA are labelled, for example at the 5′ end using polynucleotide kinase and α-^{32}P-ATP, denatured and single-stranded molecules separated and

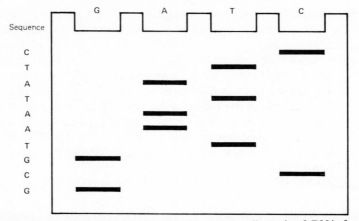

Figure 1.7 Reading the DNA sequence from an autoradiograph of DNA fragments produced by the chain terminator method. Four sets of fragments terminated with one of the four dideoxy-nucleotide triphosphate analogues are fractionated by polyacrylamide gel electrophoresis. The fragments are detected by autoradiography and the sequence is read directly from the gel.

purified by polyacrylamide gel electrophoresis. Four separate chemical treatments are given to aliquots of the DNA which result in cleavage after G, A, C + T and C. Cleavage is base-specific but occurs randomly at any of the appropriate bases in the chain. The reaction conditions are adjusted so that only partial cleavage occurs at each position, giving rise to a complete set of fragments.

In the chain termination method a short DNA primer is annealed to single-stranded DNA and the Klenow (large) fragment of DNA polymerase I is used to copy the DNA. The reactions are carried out in the presence of four deoxy-nucleoside triphosphates plus one of the four $2'3'$-dideoxy analogues (in some cases arabinofuranosyl analogues are used) in each of the four reaction mixtures. Chain termination occurs with the dideoxynucleoside triphosphates because they lack a $3'$-hydroxyl required for chain extension by DNA polymerase. The reaction conditions are adjusted so that only partial incorporation of chain terminators occurs at any site and the chains are labelled by the incorporation of an α^{32}P-labelled (or sometimes a ^{35}S-labelled) deoxynucleoside triphosphate during DNA synthesis.

Comparison of gene sequences with protein amino acid sequences has shown that the plant nucleus-cytosol system (Chapter 3) and the plastids (Chapter 4) use the 'universal' genetic code (Table 1.3). Differences are found in the genetic code used by some mitochondria (Chapter 5).

Table 1.3 The genetic code used in most organisms

First position (5′ end)	Second position				Third position (3′ end)
	U	C	A	G	
U	Phe ⎤ F Phe ⎦ Leu ⎤ L Leu ⎦	Ser ⎤ Ser ⎥ S Ser ⎥ Ser ⎦	Tyr ⎤ Y Tyr ⎦ *Term* — *Term* —	Cys ⎤ C Cys ⎦ *Term* — Trp W	U C A G
C	Leu ⎤ Leu ⎥ L Leu ⎥ Leu ⎦	Pro ⎤ Pro ⎥ P Pro ⎥ Pro ⎦	His ⎤ H His ⎦ Gln ⎤ Q Gln ⎦	Arg ⎤ Arg ⎥ R Arg ⎥ Arg ⎦	U C A G
A	Ileu ⎤ Ileu ⎥ I Ileu ⎦ *Met* M	Thr ⎤ Thr ⎥ T Thr ⎥ Thr ⎦	Asn ⎤ N Asn ⎦ Lys ⎤ K Lys ⎦	Ser ⎤ S Ser ⎦ Arg ⎤ R Arg ⎦	U C A G
G	Val ⎤ Val ⎥ V Val ⎥ Val ⎦	Ala ⎤ Ala ⎥ A Ala ⎥ Ala ⎦	Asp ⎤ D Asp ⎦ Glu ⎤ E Glu ⎦	Gly ⎤ Gly ⎥ G Gly ⎥ Gly ⎦	U C A G

Amino acids are listed according to their three-letter and single-letter abbreviations. In DNA U is replaced by T.

One version of the dideoxy sequencing method can be used to sequence the 5′ ends of RNA molecules by 'primer extension'. In this procedure a small DNA fragment, which is complementary to a sequence a little way down from the RNA 5′ end, is annealed to the RNA and acts as a primer for reverse transcription, producing a short cDNA copy of the RNA terminal nucleotides. If the reaction is performed in the presence of dideoxy analogues and a radioactive nucleoside triphosphate, then a sequence ladder of the RNA 5′ end is generated.

1.7 M13 cloning

There are several methods for obtaining single-stranded DNA for sequencing but cloning in M13 is used most frequently. M13 is a filamentous bacteriophage of *E. coli* which has a genome of single-stranded circular DNA. Insertion of foreign DNA gives a larger DNA circle which is packaged into a longer filament. The replicative form of the phage has double-stranded DNA which is used for cutting with restriction enzymes and for inserting and ligating foreign DNA. The recombinant DNA is then introduced into *E. coli* cells by transformation, and it directs bacteriophage replication. Mature M13 particles with single-stranded DNA are secreted from *E. coli*; the phage does not kill the cells but slows their growth so that they appear as clear plaques when plated out on a lawn of normal cells.

Part of the β-galactosidase gene (coding for the α-peptide) has been transferred to M 13. If the phage infects a suitable host which contains the part of the gene coding for the *w* peptide of β-galactosidase, the two DNA regions complement and the *E. coli* cells can be induced (with isopropyl-thio-β-D-galactoside) to make functional β-galactosidase enzyme. This can be detected using an analogue of the enzyme substrate (5-bromo-4-chloro-3-indolyl-β-D-galactoside) which turns blue those colonies producing β-galactosidase. DNA to be cloned is inserted into a restriction enzyme site within the α-peptide DNA sequence of M 13. This inactivates the gene and *E. coli* cells infected with phage carrying inserted DNA appear as white plaques on a blue background when plates are tested for β-galactosidase activity.

M13 is now used frequently for producing cloned single-stranded DNA for use in sequencing by the chain termination method. A number of cloning sites have been introduced into the α-peptide-coding region of M13 and complementary primers are available for initiating copy-strand synthesis of cloned DNA. Kits for M13 cloning and sequencing can now be purchased commercially.

1.8 S1 nuclease mapping

S1 nuclease mapping is a technique widely used to identify and precisely map transcribed DNA sequences. The method relies upon the specificity of S1 nuclease to digest single-stranded but not double-stranded nucleic acid. Only the transcribed DNA fragment of interest need be purified (usually a cloned fragment and preferably a single-stranded M13 clone) and this is hybridized with an RNA preparation containing the target transcript. Unhybridized regions of single-stranded DNA are degraded with S1 nuclease and the protected DNA fragment is sized by denaturing gel electrophoresis. By choosing appropriate overlapping restriction enzyme fragments, the transcribed 5′ and 3′ ends can be mapped. Very high resolution mapping (± 1 nucleotide) can be achieved by Maxam and Gilbert sequencing very small terminally-labelled DNA fragments protected from S1 digestion by hybridization with RNA. This method is also useful for mapping the junctions of intervening sequences in eukaryotic genes.

CHAPTER TWO

ORGANIZATION OF NUCLEAR DNA

2.1 The nucleus and chromatin organization

The majority of plant cell DNA is contained in a nucleus which is usually spherical or ovoid but sometimes develops lobes which greatly increase its surface area. Most higher plant nuclei are 3–20 μm across but the giant nuclei of the alga *Acetabularia* can measure up to 150 μm. The nucleus is surrounded by two membranes which are separated by a perinuclear space and perforated by pores which range from 50–100 nm in size. The effective diameter of the pores is much less than this, however, because they are partly filled with a protein complex which is presumed to be involved in transporting macromolecules through the nuclear membranes. The number and location of the pores can change quite rapidly during growth and differentiation and this reflects the dynamic nature of the nuclear envelope. It is connected to the endoplasmic reticulum and sometimes can be seen in close association with the envelope membranes of mitochondria and chloroplasts. During nuclear division the envelope is temporarily dismantled while the chromosomes segregate.

A major nuclear organelle called the nucleolus arises after mitosis at specific chromosomal locations called nucleolus organizers (Figure 2.1). They develop into prominent organelles without a limiting membrane, containing DNA and fibrils and granules of RNA and protein. They are the sites of transcription of the rRNA genes and processing and partial assembly of 80S ribosomes destined for the plant cytosol.

Nuclei contain structural proteins, such as tubulin and actin, polymerase enzymes, acidic regulatory proteins, RNA and basic proteins called histones. The ratio of DNA: histones: RNA: acidic proteins is approximately 1:1:0.1:0.6. The repeating unit of chromatin is the nucleosome, which consists of a specific aggregate of histones associated with DNA.

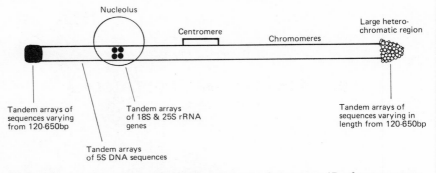

Figure 2.1 Location of repeated DNA sequences in chromosome 1R of rye as seen at pachytene of meiotic prophase. (Redrawn from Appels, 1983).

There are five histone proteins: H1, H2A, H2B, H3 and H4. Histone H1 is very rich in lysine, H2A and H2B have less lysine and H3 and H4 are rich in arginine. The primary structure of H3 and H4 is very similar in many organisms: for example the amino acid sequence of H4 is almost identical in pea and cow (De Lange *et al.*, 1969, 1973). In contrast, H2A and H2B seem more variable and there are many reports of variations in H1.

The general features of nucleosome structure appear to be similar in all eukaryotes. The nucleosome core-particle is a flattened disc, capable of separating into two halves, which measures approximately 11 nm and is composed of two molecules each of H2A, H2B, H3 and H4. The histones aggregate by hydrophobic interactions of their C-terminal regions and also by a series of alternating positive and negative charges which stabilize the structure by forming salt bridges. The basic N-terminal segments of the proteins are exposed at the surface of the particles; these interact with the phosphate groups of the DNA helix which is wound approximately twice around the outside. Histone H1 is not present in the nucleosomes but interacts with the 'linker' DNA between each bead (Figure 2.2). The length of the linker DNA seems to vary slightly in different organisms and tissues.

This model for nucleosome structure is supported by studies of their chemical composition and reconstitution *in vitro*, electron microscopy, X-ray diffraction, and sensitivity of the DNA in nucleosomes to digestion with nucleases. These latter studies tell us most about the arrangement of the DNA. Mild digestion of chromatin with deoxyribonucleases leads to the production of fragments of DNA about 170–200 base pairs long, due to cuts occurring preferentially between nucleosomes, where the DNA is least protected. Further treatment with enzymes removes all the linker DNA,

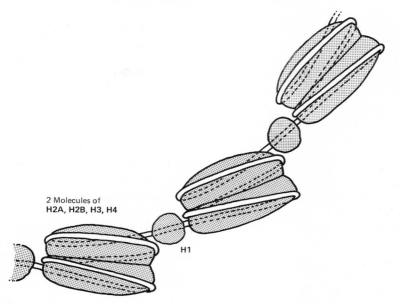

2 Molecules of
H2A, H2B, H3, H4

H1

Figure 2.2 Coiling of DNA around aggregates of histone proteins to form nucleosomes.

and produces 143-base-pair fragments associated with the core particles. The DNA wound round the core particles has a limited sensitivity to nucleases; prolonged digestion leads to cuts at intervals of approximately 10 base-pairs. It is not certain whether the DNA is wound smoothly round the nucleosome or whether it is kinked or distorted in some way.

The main result of the association of DNA with nucleosomes is to reduce the overall length of the DNA to one-seventh that of a naked molecule, resulting in a fibre approximately 10 nm thick. This condensation is much less than the several thousandfold reduction in length found in a metaphase chromosome and it is obvious that higher order structures are involved in packaging one metre or more of plant DNA into chromosomes of a single nucleus. It is thought that the nucleosomes are coiled into a 'solenoid' structure and that these are capable of condensing further. Phosphorylation of H1 is believed to play a role in chromatin condensation during mitosis. Modification of other histones may also affect chromatin structure but we do not have a clear picture of how this is brought about. Neither do we have a clear understanding of the molecular basis of the cytological distinction between the differently-staining regions of euchromatin and heterochromatin (Figure 2.1), although it is known that the latter has a high proportion of repeated sequences which are not transcribed into RNA. If the general interactions between DNA and histones are simply electrostatic, then formation and spacing of nucleosomes should be independent of DNA sequence. However, there is some evidence, from work with animals, that satellite DNAs, which have repeating sequences at regular intervals, have a non-random distribution of nucleosomes. This suggests that the actual sequence may influence nucleosome spacing or arrangement.

There is no general agreement about the arrangement of DNA in nucleosomes while it is being copied by polymerase enzymes. It is possible that during transcription or replication the nucleosomes dissociate into two half-nucleosomes. Indeed, there is some evidence that newly-synthesized DNA (see section 2.2) is susceptible to nuclease and may not be in the normal nucleosome configuration during replication.

2.2 DNA replication

Most actively dividing plant cells complete a cycle of growth and cell division in 15–40 hours depending on the species and the temperature. DNA replication and histone synthesis, which lasts for 7–11 hours, is confined to part of the interphase cycle, known as the S phase.

DNA synthesis has been studied by labelling plants with radioactive precursors of DNA and at intervals extracting and preparing the DNA for autoradiography of individual molecules. The results show that each molecule is replicated at many hundreds of points along its length. Replication proceeds in both directions from each origin of replication; each segment of DNA thus synthesized is called a replicon. Various researchers have shown that the replicons in higher plants are generally from 20–30 μm or 60–90 kilobase pairs long. The replication forks, which number from 5 000–60 000 per diploid genome, move at about 10 μm per hour (Van't Hof and Bjerknes, 1979). There are from 2–25 families of replicons in different higher plants; each replicon in a family undergoes DNA synthesis at the same time during S phase but separate families are active at different times.

DNA replication in plants is likely to involve the soluble α- and γ-DNA polymerases; the chromatin-bound β-polymerase is probably involved in DNA repair. Our understanding of the biochemistry of DNA replication in higher plants is rather poor, however, and there are several gaps in our knowledge. The following speculative model (see Figure 2.3) is based on

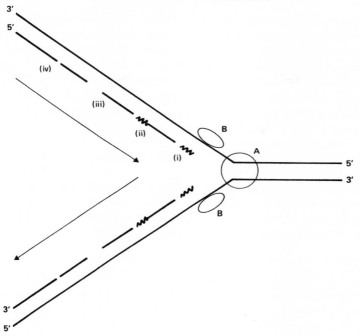

Figure 2.3 A model of events at the replication fork during DNA synthesis in plants. Step (i), synthesis of RNA primer; (ii), chain extension; (iii), primer excision; (iv), gap-filling and ligation. A is 'unwinding enzyme', B is the postulated single-stranded DNA-binding protein.

knowledge gained from studies on animals and lower eukaryotes (Bryant, 1982). (1) DNA synthesis is initiated at a replication fork by nicking and unwinding of the DNA by a topoisomerase ('unwinding enzyme'). (2) The separated DNA strands are bound by a 'DNA-binding enzyme'. (3) A short RNA primer is synthesized in the $5' \rightarrow 3'$ direction, copying each DNA strand in opposite directions. (4) DNA polymerase elongates the primer in the $5' \rightarrow 3'$ direction to produce nascent DNA molecules called Okazaki fragments (named after one of the researchers who first discovered them). Appropriate fragments, about 200 nucleotides long, have been demonstrated in soybean and the slime mould *Physarum*. (5) RNA primers are then removed from the Okazaki fragments by a nuclease. (6) The resulting gaps in the DNA are repaired by DNA polymerase. (7) The adjacent DNA segments are ligated together to produce a new daughter strand. (8) New RNA primers and Okazaki fragments are synthesized as the replication fork progresses. (9) Eventually, newly-synthesized DNA strands from adjacent replicons are ligated together.

It is possible that the 200-nucleotide Okazaki fragments, which are shorter than those found in prokaryotes, are related to the arrangement of DNA in the nucleosomes. Studies on the replication of SV40 and polyoma viruses in animal cells show that the initiation of DNA synthesis begins at short partial tandem repeat sequences: 5′ TTAACGTTAA 3′. Such sequences have a twofold rotational symmetry (the sequence is the same when each strand is read in the 5′ to 3′ direction) and could form hairpin loops in the DNA which might act as signals for initiation of replication.

After DNA replication in higher plants, up to 25% of the cytosine residues are methylated by a DNA methylase that uses S-adenosyl methionine as methyl donor. Sites at which C is commonly methylated are GC, CG, CAG, GTC.

2.3 Nuclear DNA amounts and the C-value paradox

Plant nuclear DNA amount has been estimated by direct chemical analysis, allowing for the number of cells in the sample, and by micro-densitometry of individual Feulgen-stained nuclei. These values are normally expressed as the 1C or 2C level of DNA, or the amount present in a haploid or diploid cell before S phase (Table 2.1). The nuclear DNA content in different

Table 2.1 Range of 1C nuclear DNA values in higher plants compared to a virus, a chloroplast, bacterium and humans

	DNA content (grams per genome)	Number of base pairs in genome	Molecular weight
Cauliflower mosaic virus	0.84×10^{-17}	8 024	5.1×10^{6}
Pea chloroplast	1.3×10^{-16}	1.24×10^{5}	80×10^{6}
Escherichia coli	4.0×10^{-15}	3.9×10^{6}	2.5×10^{9}
Mung bean	0.48×10^{-12}		
Soybean	1.2×10^{-12}		
Parsley	1.9×10^{-12}		
Tobacco	2.0×10^{-12}		
Pea	4.5×10^{-12}	1 picogram $(10^{-12}\,g)$	
Wheat	5.0×10^{-12}	is approximately	
Rye	7.9×10^{-12}	10^{9} base pairs or 6.4×10^{11} daltons	
Oats	4.3×10^{-12}		
Barley	5.5×10^{-12}		
Maize	6.2×10^{-12}		
Humans	6×10^{-12}		

higher plants ranges from 0.5 to over 200 picograms (Benett and Smith, 1976). Comparison with values from other organism shows that pea, for example, contains 1000 times as much DNA as *E. coli*, which probably contains about 4000 genes (Table 2.1). Thus, if all the pea DNA was expressed in the conventional sense, it would represent four million genes. While it is difficult to estimate the number of genes necessary to account for the differences in biological complexity between pea and *E. coli*, a value as high as this seems extraordinary, especially as it would mean that many plants would have 100 times as many coding sequences as humans, which only have about 6×10^{-12} g DNA and yet are generally regarded as more complex. In addition, organisms with such a large number of essential genes would be prone to a very high number of mutations affecting the phenotype. Further comparisons of the DNA content of plant nuclei show that there is, for example, a threefold variation between species of the genus *Lathyrus* and a tenfold variation in the genus *Crepis* (Flavell, 1982). It seems totally unreasonable to conclude that such large variations in DNA content underlie the genetic differences between closely related species and so clearly there must be some other explanation for the large differences in DNA content.

Estimates of the haploid DNA content of higher plant nuclei have also been made by studying the kinetics of DNA renaturation (see section 2.4). For at least some plants such as pea, tobacco and soybean, the results are in good agreement with chemical and cytological measurements. However, the kinetic analysis also shows that a large proportion of the DNA is composed of highly repetitive sequences and is unlikely to represent genes coding for proteins. A reasonable estimate of the number of genes that do code for proteins in higher plants, based on genetic and biochemical considerations, is from 40–100 000. Assuming that an average gene is equivalent to 1000 base pairs, this represents 40–100 million base pairs. For a plant such as pea (Table 2.1) this works out at about 1–2% of the total haploid DNA content.

2.4 Organization of DNA sequences

The complexity of plant DNA has been studied in a number of ways but the kinetic measurements of DNA reassociation (discussed by Flavell, 1982) give the best overall picture of genome organization and will be outlined first.

When double-stranded DNA is heated above the T_m (melting temperature) or treated with alkali, the hydrogen bonding is disrupted and the strands separate. Conversely, when the temperature is lowered or the pH restored to near-neutrality, complementary strands are capable of re-associating. The rate of reassociation is dependent on temperature, salt concentration and, for complex sequences, on the length of the DNA fragments. If these parameters are controlled then the reassociation rate is proportion to the concentration of reacting sequences and time and follows second-order kinetics. The $C_0t^{1/2}$ value (where $C_0 =$ concentration of double-stranded DNA and t is time), when half the DNA is renatured, is inversely proportional to the reaction rate and for simple genomes with no repetitive DNA is related to genome size. The reassociation rate is monitored either by measuring the hypochromicity at 260 nm with a spectrophotometer, by separating single-stranded and double-stranded DNA by chromatography on hydroxapatite (HAP) or by S1 nuclease treatment, which specifically digests single-stranded DNA. With pea DNA only about 15% of the genome behaves as if it is present in one or a few copies (Figure 2.4, Murray *et al.*, 1978). Approximately 85% of the DNA reassociates much more rapidly and is composed of many families of sequences repeated thousands of times. Similar results have been found

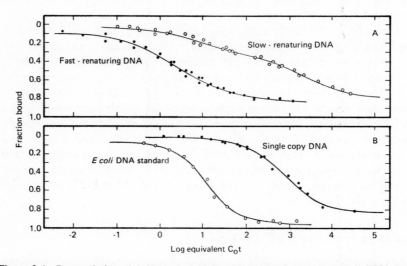

Figure 2.4 Reassociation of single-stranded DNA from mung bean leaves. In A, DNA was renatured to $C_0t = 5$ and separated by hydroxyapatite chromatography into fast (●) and slow (○) renaturing fractions. The reassociation rate of these two fractions was then studied separately by allowing each to reassociate with ^{14}C-labelled tracer DNA. In B, a fraction enriched in single copy DNA (●) (not bound to hydroxyapatite after renaturation to $C_0t = 300$) was mixed with a trace of ^{14}C-labelled DNA and the reassociation rate compared to that of *E. coli* DNA (○). (Redrawn from Murray *et al.*, 1978.)

with many higher plants. In general, the proportion of 'single copy' DNA (i.e. the fraction of the genome which is present in one or a few copies) ranges from 20–40%. However, this varies with genome size because there is a tendency for the proportion of repetitive DNA to increase in plants with larger genomes (Flavell, 1982). DNA–RNA hybridization experiments show that only a small proportion of the 'single-copy' DNA is transcribed into mRNA. In tobacco leaves, for example, the polyribosomes contain about 27 000 different transcripts (Goldberg *et al.*, 1978) and the total number of genes transcribed in all tobacco tissues is approximately

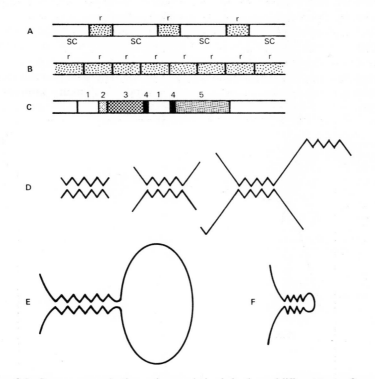

Figure 2.5 Sequence organization and reassociation behaviour of different types of repeated DNA. *A*, interspersion of a repeated sequence between single-copy DNA. *B*, tandemly-repeated DNA sequences. *C*, repeated DNA interspersed with different repeats. *D*, detecting repeated sequences interspersed between single-copy DNA by studying the effect of fragment size on the proportion of DNA present as duplexes after reassociation of repeated sequences. *E*, intra-strand renaturation of an inverted repeat sequence separated by single-copy DNA. *F*, intra-strand renaturation of adjacent reverse-repeat sequences. The regions drawn as zigzags indicate base-paired sequences.

60 000. The structure and expression of single-copy genes is discussed in Chapter 3.

Some repeated DNA sequences have been shown to be interspersed between single copy DNA sequences (Figure 2.5). Evidence for interspersion comes from studies on the reassociation of DNA fragments of different lengths. When DNA is renatured to appropriate C_0t values, only the repeated sequences form duplexes. If the DNA segments used in the reaction are relatively short, only the duplex regions bind to hydroxyapatite and single-stranded DNA passes straight through the column (Figure 2.5 and Figure 2.6). However, when the experiment is repeated with longer DNA fragments, single-stranded tails are found attached to the duplex regions. These also bind to hydroxyapatite (Figure 2.5 and Figure 2.6) but can be shown to be single-stranded because they are sensitive to S1 nuclease. In some plants the interspersed repeats are, on average, less than 1000 base pairs long and are located between single copy regions of the order of 1–2000 base pairs long. However, a range of values

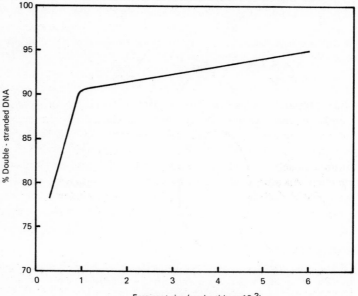

Figure 2.6 Detecting repeated DNA sequences interspersed between single-copy DNA. Single-stranded DNA of different lengths is renatured to a C_0t value which allows repeated sequences to reassociate and the DNA fraction attached to double-stranded DNA measured by hydroxyapatite chromatography.

for the lengths of these sequences has been found by researchers working with different plants.

Inverted or reversed repeat sequences have also been detected by DNA reassociation; these sequences can form hairpin loops within one DNA strand (Figure 2.5). This occurs at extremely low C_0t values and can be measured by HAP chromatography, S1 nuclease digestion and electron microscopy. The reverse-repeats range from 50–1000 base pairs long. In wheat, about 20% of the genome contains reverse-repeats which are interspersed with less repetitive DNA sequences between 300–1000 base pairs long (Bazetoux et al., 1978). Adjacent reverse-repeat sequences, of the general form discussed in relation to initiation of DNA replication, may also occur. However, short sequences of this type (Figure 2.5) are not detected by HAP chromatography and electron microscopy but are only revealed by sequencing. Other types of repeated sequences occur in large blocks, either as tandem repeats of one or a few sequences (such as the rRNA genes discussed below) or as a more varied permutation of several different repeating regions (Figure 2.5). Sometimes these sequences can be physically separated as satellite DNAs (section 2.5). Many repeated sequences have now been cloned, mapped by restriction enzymes and the DNA sequence determined. Some of them are quite long but in many cases appear to have evolved from short ancestral sequences which have been amplified and subsequently diverged by mutation. Many have been located on chromosomes by hybridization of cloned nucleic acid probes to cytological preparations in situ. The location of some repetitive sequences at the ends of chromosome R1 from rye is shown in Figure 2.1. The distribution of repetitive DNA is quite variable; it can be located in telomeric, centromeric or interstitial heterochromatin sites (Flavell, 1983). In addition, closely related sequences are often found at thousands of separate locations among chromosomes and there is strong circumstantial evidence that they can duplicate and change position during evolution. It seems probable that transposable elements are involved in this process.

2.5 Satellite DNAs

Satellite DNAs were first detected as distinct subfractions of the genome which had sufficiently different properties from the bulk of the 'main band' DNA to enable them to be separated by isopycnic (buoyant density) centrifugation in CsCl solutions. Some of the first animal satellites to be studied were found to have simple repeating sequences associated with centromeres or other heterochromatic regions of chromosomes. However, the term 'satellite' can be applied to any DNA sequence physically separated by isopycnic centrifugation and does not imply a particular origin or function. The actual density of DNA depends on the nature of the salt, whether the DNA is single- or double-stranded, and can be influenced by pH and the addition of heavy metals or DNA-binding dyes and drugs. In CsCl the main factor determining buoyant density is the base composition, whereas in Cs_2SO_4, actinomycin D and Hg^{2+} bind to, and specifically alter the density of, GC-rich DNA, and Ag^+ binds preferentially to AT-rich DNA. Many satellites have been described in higher plants (Ingle *et al.*, 1973). They consist of long stretches of DNA

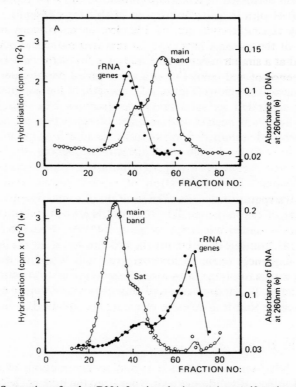

Figure 2.7 Separation of melon DNA fractions by isopycnic centrifugation. Melon DNA was centrifuged in CsCl, separated into main band and satellite DNA fractions and the rRNA genes located by hybridization with radioactive rRNA (*A*). Centrifugation of the DNA in Cs_2SO_4-actinomycin gradients increases the separation between rRNA genes and satellite DNA, due to preferential binding of actinomycin by the rRNA sequences (*B*). (Redrawn from Hemleben *et al.*, 1977.)

containing repeated sequences which are different from the bulk of the genome. They can represent either a large or a small proportion of the total DNA and the detection of minor satellites depends on the sensitivity of the method used. A variety of satellites has been characterized and their sequences and chromosomal location determined by *in situ* hybridization. The repeat length of plant satellites ranges from six to several hundreds of base pairs. Occasionally, organelle genomes have been mistakenly assumed to be nuclear satellite DNAs.

The detection and separation of two plant satellite DNAs is illustrated in Figure 2.7. In CsCl, melon DNA separates into a main band plus a major satellite (part of this may represent organelle DNA). Hybridization of each DNA fraction with radioactive 25S and 18S rRNA pinpoints the position of the nucleolar rRNA genes, which have a higher GC content than the main part of the genome in many plants (Figure 2.7*A*), and with melon DNA band at a similar position to the satellite. Centrifugation in Cs_2SO_4 in the presence of actinomycin D shows that the density, relative to the main band, of the GC-rich satellite and rRNA genes is decreased by binding the drug. However, the two sequences do not bind actinomycin D equally well and this enables a substantial purification of the rRNA genes to be achieved (Figure 2.7*B*).

2.6 Nucleolar genes for rRNA

The general structure and organization of the genes coding for rRNA was deduced before they were isolated. DNA–RNA saturation hybridization experiments generally show that from 0.1–1% of plant DNA is homologous to 18S and 25S rRNA from the 80S ribosomes of the cytosol. The two gene sequences occur in approximately equal numbers. For most plants this works out to be from a few hundred to several thousand genes per diploid cell. There is substantial variation in rRNA gene number even within a species. Studies on rRNA synthesis *in vivo* (reviewed by Grierson, 1982) show that rRNA is first transcribed as a polycistronic precursor RNA which contains sequences for 18S and 25S rRNA. The size of the precursor is larger than that required to code for these sequences alone and it contains other transcribed (spacer) regions of the DNA (Chapter 3). The rRNA genes were originally shown to be associated with nucleoli by studies with *Xenopus* mutants which lack a nucleolus and fail to make ribosomes because the rRNA genes are deleted. More recently, radioactive plant rRNA and cloned DNA sequences coding for rRNA have been hybridized to chromosomes *in situ* and the location of the genes in plant nucleoli confirmed (Figure 2.1). Many plant rRNA gene sequences have now been purified, cloned, and restriction enzyme maps determined. The repeat unit ranges from 8–11 kilobases in various plants and hundreds or even thousands of repeats are arranged in tandem. The organization of the soybean rRNA genes is shown in Figure 2.8. The rRNA sequences were located by hybridization to several different sets of restriction enzyme fragments. The 5′ ends were mapped with rRNA fragments specifically labelled *in vitro* at the 5′ end by polynucleotide kinase and α-^{32}P-ATP. Similar gene arrangements are found in other higher plants. There are differences in repeat length and sequence but this is due to variation in the 'spacer' DNA. Three different repeat units are present in wheat, with lengths of 9.0, 9.15 and 9.45 kilobases (a kilobase is 1000 bases or base pairs). The smaller two sequences are located mainly on chromosome 6B and the largest sequence is found on chromosome 1B (Appels, 1983). The 9.0 kilobase variant of wheat rRNA genes has been shown by restriction enzyme digestion and sequencing to have an internal repeating structure of 130–150 base pairs in the spacer DNA. This repeating sequence shows some variation through the spacer region but is not found in repeated sequences elsewhere in the wheat genome (Appels, 1983).

Multiple rRNA genes are a physiological necessity for plants. A typical higher plant cell contains several million ribosomes. If the cell divides every 15 hours it has to make several hundred thousand ribosomes per hour. At

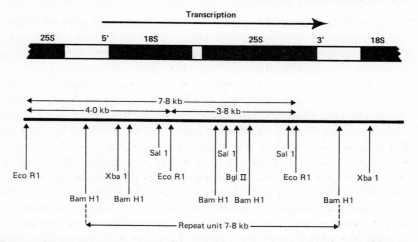

Figure 2.8 Location of restriction enzyme sites in the tandemly-repeated nucleolar rRNA genes in soybean. The orientation (5′, 3′) is with respect to the strand with the *same* sequence as rRNA.

the temperatures at which plants often grow it probably takes several minutes for an RNA polymerase enzyme to transcribe a complete rRNA gene sequence. Therefore, in order to explain the observed rate of ribosome production, one has to predict the existence of multiple sites of rRNA synthesis. This is achieved partly by having multiple genes and partly also by transcribing each gene many times simultaneously, as has been observed by electron microscopy of spread nucleolar contents from *Acetabularia* (see Chapter 3). In some plant cells, some of the multiple rRNA genes are condensed in heterochromatin and appear not to be expressed. It is not clear whether these genes are always surplus to requirements or whether this only occurs at certain stages of the life cycle. In wheat, some rRNA genes are methylated at the internal C in CCGG sequences. Genes that are methylated at all these sites are present in condensed chromatin and are transcriptionally inactive.

Multiple 5S rRNA genes in tandem arrays are also found in plants, unlinked to the 25S and 18S rRNA genes (Figure 2.9 and Chapter 3). In the Chinese Spring variety of wheat, two length variants have been detected, of 410 and 500 base-pairs, which differ mainly in the spacer DNA. The longer sequence has also been shown to contain a 15 base-pair duplication in the 5S coding region and these genes may not function (Gerlach and Dyer, 1980). The 70 base-pair region immediately upstream from the coding sequence is strongly conserved in the two genes and contains 5′ ATAAG 3′, which is also found in *Drosophila* 5S genes, and is believed to be important in regulating transcription. The AT-rich region after the 3′ end of the

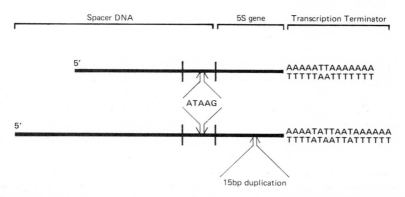

Figure 2.9 Organization of two types of 5S rRNA genes in wheat. (Redrawn from Flavell, 1982).

coding sequence is believed to be a termination signal for RNA polymerase (Figure 2.9).

Many studies indicate that spacer sequences show much more variation than coding regions and the question arises as to how multiple coding sequences remain similar during the course of evolution. There are several ways in which this might be explained. One proposal (Appels, 1983) is that during transcription the RNA sequence displaces one of the DNA strands in the double helix, and that this free DNA invades a neighbouring strand to form a heteroduplex. Any mismatched sequences are then excised and repaired, thus leading to a 'gene conversion' phenomenon. Clearly, non-transcribed sequences would be unable to participate in this type of reaction and thus might diverge more rapidly.

2.7 Function of repeated DNA sequences

It is very probable that some repetitive DNA sequences are control regions involved in regulation of transcription. Sequences of this type would be expected to be interspersed in single copy DNA and in transcribed repeated DNA such as the rRNA genes. Relatively short transcription initiation and termination sites would be expected (see Chapter 3) but other sequences may be involved in long-range interactions and repression or activation of DNA regions. Massed repeating sequences in centromeres and other sites could be involved in chromosome pairing or attachment to features of the nuclear skeleton. However, species from the same genus often differ in their content of repeated and satellite DNAs and it is probable that not all repeated sequences have a specific function.

2.8 Genome variation

Although we are used to thinking of the genome as being constant, with the exception of rare mutations, it has been shown that the environment can affect the plant genotype as well as the phenotype. Experiments with flax show that when the plastic form (P1) is grown in a heated greenhouse under certain fertilizer regimes and soil pH, it can change to a large (L) or small (S) genotroph according to the balance of nutrients supplied. The L and S forms are stable and the change in size is inherited. Inherited alterations in nuclear DNA amount and in the numbers of 25S + 18S and 5S rRNA genes have been shown to be associated with these effects. It has been suggested that these environmentally-induced heritable changes might be caused by unequal duplication of some DNA sequences, unequal crossing over, or some extrachromosomal event leading to sequence duplication and rearrangement (Cullis, 1983).

2.9 Transposable elements

Transposable elements, or insertion elements, are DNA sequences which can be inserted into many different sites in chromosomes. Some have the capacity to move around the genome, either on their own or under the influence of other sequences. For this reason they are sometimes called jumping genes. Their movements can be detected genetically, because when a mobile element is inserted in a structural gene, the gene may be inactivated.

Transposable elements in maize were detected and characterized by McLintock, before the structure of DNA was known (McLintock, 1945, 1951). Similar genetic elements were subsequently discovered and characterized in bacteria, yeast, invertebrates and vertebrates and many have now been sequenced. Nearly 40 years after her discovery of mobile genetic elements in maize, McLintock was awarded the Nobel Prize for Physiology and Medicine.

One type of maize insertion element, known as *Dissociator* (*Ds*) cannot be transposed by itself but requires another element, known as *Activator* (*Ac*). Phenotypic effects of these elements are detected when they disrupt particular genes such as at the 'shrunken' locus in maize, which affects the sucrose synthase enzyme in the endosperm. Starlinger and his colleagues (Döring *et al.*, 1984) have sequenced a 4.2-kilobase region of DNA, believed to correspond to *Ds*, which forms part of a larger inserted sequence at the 'shrunken' locus. The *Ds* sequence consists of two identical 2040 base-pair segments, one inserted into the other in the reverse orientation. This *Ds* sequence has an 11 base-pair inverted repeat 5'TAGGGATGAAA 3' at each end and seems to create a short direct repeat at the insertion site in the chromosome. There are many copies of the *Ds* sequence throughout the genome of maize. Other researchers have isolated different *Ds* sequences. One, found in an alcohol dehydrogenase gene, is 450 base-pairs long and another is 1300 base-pairs long. Another maize transposable element, *Cin* 1, which is 691 base-pairs long, has a 6 base-pair inverted repeat 5' TGTTGG 3' which is different from that found in *Ds*. This *Cin* 1 element is not flanked by a direct repeat, but two other *Cin* 1 alleles are (Shepherd *et al.*, 1984). The *Cin* 1 sequence contains a number of interesting features including RNA polymerase binding sites, poly-A addition signals and ATG codons (see Chapter 3). It is possible that some of these elements are transcribed and translated. The *Ds* element contains an ATG initiation codon for protein synthesis which is followed by a region (bases 3373–3981) which could code for a protein of 203 amino acids. Similarly, the *Cin* 1 sequence could encode a polypeptide of 69 amino acids.

Some of the features of the maize transposable elements are similar both to insertion sequences in *Drosophila* and to retroviruses, which are animal RNA viruses copied into DNA by reverse transcriptase and then integrated into chromosomes (Döring *et al.*, 1984; Shepherd *et al.*, 1984). The characterization of insertion elements is thus beginning to provide a molecular explanation of McLintock's genetic observations. It also suggests ways in which sequences might move around the genome and could be relevant to the origin and development of repetitive DNA. Furthermore, it raises intriguing questions about the relationship between chromosomal DNA sequences and viruses. Finally, although we do not yet fully understand how the transposable elements operate at the molecular level, the essential features of the sequence may provide a vehicle for genetic engineering (Chapter 10).

CHAPTER THREE

STRUCTURE AND EXPRESSION
OF NUCLEAR GENES

3.1 RNA polymerases

Plant nuclear DNA sequences are transcribed by three separate forms of polymerase enzyme referred to as RNA polymerases I, II and III. They are all multisubunit proteins which require a DNA template, nucleotide triphosphates and manganese or magnesium ions for activity. The enzymes are structurally and functionally distinct and resemble closely the RNA polymerases found in animal cell nuclei. The subunit compositions have been determined for enzymes purified from a number of different plants but we have very little idea of the function of individual polypeptides or the role that specific factors might play in altering their transcriptional activity.

RNA polymerase I is located in the nucleolus where it transcribes the genes for rRNA. The enzyme has been reported to consist of 6–10 subunits with molecular weights ranging from 8000–185000. RNA polymerase I is active at low ionic strength and is resistant to the fungal toxin α-amanitin. These properties enable it to be distinguished from the other two nuclear enzymes.

RNA polymerase II is located in the nucleoplasm. It consists of from 8–14 subunits which range in molecular weight from 14000–220000. The subunit composition of the enzyme has been found to be different when extracted from dormant embryos compared to actively growing tissue. It is not clear whether this has regulatory significance or whether it simply results from proteolysis during enzyme purification. RNA polymerase II is important in the transcription of genes coding for proteins. It is sensitive to low concentrations of α-amanitin and 50% of the enzyme activity can be inhibited by $0.01–0.05\,\mu\mathrm{g\,ml^{-1}}$ of the drug.

RNA polymerase III is reported to consist of 8–14 subunits with molecular weight from 16000–155000. The enzyme is believed to be responsible for the transcription of low molecular weight sequences in the nucleoplasm including the genes for 5S rRNA and tRNA. It can be distinguished from the other nuclear RNA polymerases by virtue of its sensitivity to high concentrations (50–1000 μg ml^{-1} for 50% inhibition) of α-amanitin.

3.2 Transcription and processing of rRNA and tRNA

In non-photosynthetic tissue the vast majority of the ribosomes (several million in each cell) are of the 80S type present in the plant cytosol. These ribosomes, which are responsible for the translation of mRNA molecules transcribed from nuclear DNA, are either free in the cytoplasm or attached to endoplasmic reticulum membranes. They contain the 25S, 5.8S and 18S rRNA components which account for approximately 70% of the total cellular RNA (Table 4.5) and are assembled in the nucleolus. Plant cells also contain much smaller quantities of 78S mitochondrial and 70S plastid ribosomes which are synthesized within these organelles and are responsible for the separate translation of organelle-encoded mRNAs (Chapters 4 and 5). In leaves, however, where plastid protein synthesis represents a very high proportion of the total, up to 50% of the ribosomes are of the 70S type.

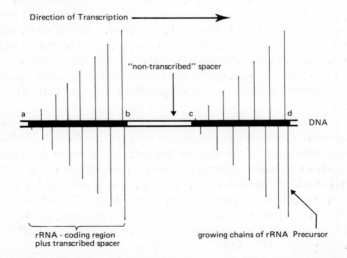

Figure 3.1 Transcription of the rRNA repeating units in nucleolar DNA. (Redrawn from Grierson, 1982.)

Since rRNA is synthesized at very high rates in growing tissues its production is easy to study. The main features of the process have been worked out by feeding radioactive precursors of RNA such as ^{32}P-phosphate or ^{3}H-uridine to plant cells or tissues and determing the sub-cellular location, properties and metabolism of the newly-synthesized RNA (reviewed by Grierson, 1982). Many hundreds of rRNA genes are massed in tandem arrays in the nucleoli of plant cells (Chapter 2). They have a general repeating structure consisting of transcribed and non-transcribed regions, illustrated in Figure 3.1. The 'non-transcribed spacer' (an unfortunate misnomer, see below) varies considerably in length and composition in different organisms and even within a species; it has been shown to consist of repeating sequences (Chapter 2). Electron microscopy of the spread nucleolar contents from many organisms, including the alga *Acetabularia*, shows there are multiple RNA polymerase I molecules transcribing each tandem repeat of the rRNA genes at a number of sites simultaneously, as depicted in Figure 3.1. The exact nature of the initiation and termination sequences for rRNA transcription by plant RNA polymerases I is not known. The initial transcripts contain one 25S, 5.8S and 18S rRNA sequence (equivalent to about 5660 nucleotides) and some non-conserved RNA sequences. The rRNA precursors vary in molecular weight in different plant species, from 2.1–3.4 million (between 6500 and 10 000 nucleotides). The differences are thought to be due to variability in

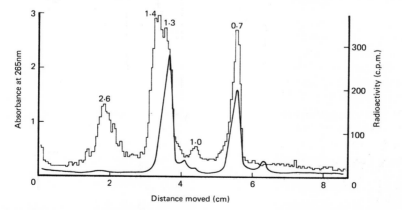

Figure 3.2 Polyacrylamide gel electrophoresis of rRNA precursors and processing intermediates from mung bean roots. Seedlings were pulse-labelled with ^{32}P-phosphate for 1 hour and the RNA extracted and fractionated. The histogram shows the distribution of radioactivity; the smooth curve shows the location of the rRNAs. The approximate molecular weights of the RNAs are given in millions (see Figure 3.3). (Redrawn from Grierson, 1977.)

the length of 'transcribed spacer' surrounding the rRNA coding sequences but may also be due to introns within the coding regions, as found for *Chlamydomonas* chloroplast rRNA (Chapter 4). It takes several minutes to complete the synthesis of each rRNA precursor molecule, which exists separately for some minutes in the nucleolus (Figure 3.2). During subsequent 'maturation' the precursor undergoes chemical modification and nucleolytic cleavage carried out by processing enzymes. Approximately 100 sites are methylated (mainly on the ribose) and smaller processing intermediates are generated in the molecular weight range 1.45–1.4 million and 1.0–0.9 million (Figures 3.2 and 3.3). These are finally cleaved to remove terminal and intervening RNA sequences, leading to the production of mature rRNAs which are combined with about 100 ribosomal proteins and assembled into 60S and 40S ribosome subunits in the nucleolus and nucleoplasm (Figure 3.3). The 5.8S rRNA arises as part of the 25S rRNA precursor and remains hydrogen-bonded to it after the removal of intervening sequences. In *Tetrahymena* it has been shown that the introns which are transcribed as part of the rRNA precursor contain nucleotides at the borders which, in conjunction with a guanosine cofactor, catalyse a cleavage and ligation reaction resulting in the removal of the introns. Although the reaction *in vitro* does not require any proteins, it is possible that it is catalysed by enzymes *in vivo* (Filipowicz and Gross, 1984).

In some organisms it seems probable that the initiation sites for rRNA precursor transcription are in altered positions in different genes, thus giving rise to transcripts with different lengths of non-conserved RNA. Since the presence of introns is not essential for rRNA synthesis, and since the amount of transcribed non-conserved RNA varies such a lot, it seems probable that these sequences do not all perform an essential function during rRNA production. However, some of the extra nucleotide sequences in rRNA precursor molecules probably function in a transient manner to induce formation of the correct secondary and tertiary structure during ribosome assembly. The hundred or so methylation events that occur are all in regions of the rRNA molecules which are highly conserved in different species, suggesting that methylation plays a fundamental role in ribosome structure. The nucleotide sequence of the 18S rRNA from soybean shows more than 75% homology with those from *Xenopus* and yeast and the proposed secondary-structure models are very similar. This indicates that there are many features common to all eukaryotic rRNAs.

Electron microscopy of nucleolar contents from animal cells shows transcription does occur in the 'non-transcribed spacer' regions and this is

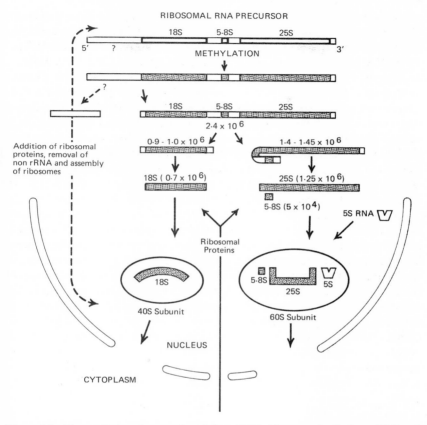

Figure 3.3 The synthesis and processing of plant rRNA. The genes for the large rRNAs of the 80S ribosomes are transcribed in the nucleolus as a polycistronic precursor. The transcript is methylated, cleaved by processing enzymes and the rRNAs assembled into ribosomal subunits with proteins synthesized in the cytoplasm. The 5S rRNA sequence is transcribed from a separate chromosomal location. (Redrawn from Grierson, 1982.)

probably the same in plants. The results from DNA sequencing indicate that this may take place at positions where the sequence resembles that of the true initiation site for the rRNA precursor. One explanation of these observations is that the spacer region amasses RNA polymerase I molecules by forming transcription complexes, thus ensuring an abundant supply of enzymes is available for transcription of the nearby rRNA coding regions.

In higher plants the 5S rRNA genes, which are also present in multiple copies, are not physically linked to the 25S and 18S genes but are located elsewhere on the chromosome (Chapter 2). This differs from the arrangement in bacteria, some lower eukaryotes and chloroplasts (Chapter 4). The 5S rRNA genes are transcribed by RNA polymerase III. Very little is known about the structure or processing of 5S precursor RNAs in plants.

Multiple tRNA genes are also transcribed as precursor RNAs by RNA polymerase III. Details of the processing have been worked out for bacteria and yeast. One of the processing enzymes, ribonuclease P, contains a tightly-bound RNA molecule which functions as the catalytic site for cleavage of the precursor (Guerrier-Takada *et al.*, 1983). Some tRNA precursors also contain introns which are removed and the tRNA halves ligated together during processing. The mechanism for this series of reactions in plants has been described by Filipowicz and Gross (1984).

In some cases the terminal CCA of the tRNA is not coded genetically and is added after synthesis. During processing the molecules of tRNA undergo extensive modification to produce the rare bases which have long been known to be a feature of tRNA structure. Notable among these changes is the frequent chemical substitution of a purine next to the anticodon of the tRNA to produce a base which is similar in structure to the naturally-occurring group of plant hormones known as cytokinins. The presence of modified purines next to the anticodon seems to increase the binding of tRNAs to ribosomes during protein synthesis. It is thought that their occurrence in tRNAs is not related directly to the function of plant hormones.

3.3 Transcription, processing and translation of mRNA

Nuclear genes coding for proteins are transcribed by RNA polymerase II. The mRNAs, which represent 2–4% of the cellular RNA, appear to be monocistronic (that is each mRNA codes for only one polypeptide) and range from a few hundred to several thousand nucleotides long, depending on the size of the proteins for which they code. In addition to the coding region, the mRNAs contain 5′ and 3′ sequences of untranslated nucleotides of variable length which have special features (Figure 3.4).

Chemical analysis of some plant mRNAs has shown that they contain a 5′ cap structure related to that found in animals. This is an unusual 5′ → 5′ triphosphate linkage involving a 7-methylguanine residue which is added after transcription. The cap is not essential for mRNA translation *in vitro* but is thought to function by binding a specific protein which may regulate initiation of protein synthesis. Since the translation in cell-free systems of many plant mRNAs from the cytosol is inhibited by cap analogues such as 7-methylguanine-5'-monophosphate, it seems probable that the majority are capped. Translation of chloroplast mRNAs on the other hand is unaffected by cap analogues and they are assumed to have no special 5' cap structure. However, they do have a short nucleotide sequence in the 5' untranslated region of the mRNA which is capable of hydrogen-bonding with a complementary sequence at the 3' end of the 16S rRNA of the 70S chloroplast ribosomes (Chapter 4). No such sequence has been detected in nuclear mRNAs which would allow hydrogen bonding with the 18S rRNA of the 80S ribosomes.

The 3' ends of many plant mRNAs have been shown to contain poly-A (poly-adenylic acid) regions up to 200 nucleotides long. These are not coded genetically but are added after synthesis by a poly A-polymerase in the nucleus. Quantitative studies have shown that only about half of the cytosol mRNA molecules bind to oligo(dT)-cellulose or poly(U) sepharose and therefore contain long poly-A sequences. The results of *in vitro* protein synthesis experiments with the two types of mRNA indicate that the same coding sequences are present in the poly A-plus and poly A-minus fractions. It is probable, however, that mRNAs with short runs of A are included in the polyA-minus fractions. There is some evidence that the stability of mRNA is related to the length of the poly-A sequence and it is possible that 'older' mRNAs have shorter 3' poly-A regions.

Pulse labelling experiments have shown that much of the heterogeneous nuclear RNA in plants is about 50% larger than the mRNA associated with polyribosomes (Grierson, 1982). It is rapidly polyadenylated after transcription and is presumed to undergo further processing, including the removal of intervening sequences (see below), before it enters the cytoplasm. The results of DNA–RNA hybridization experiments suggest that about 25% of the sequences present as heterogeneous nuclear RNA (hnRNA) in plants actually reach the cytoplasm. However, unlike the situation for rRNA, there have been very few direct studies of mRNA synthesis and processing in plants and in general the picture has been pieced together by comparing mRNA structure with gene structure (Figure 3.4) and by analogy with studies on mRNA synthesis in animal cells.

In some specialized plant cells the major part of protein synthesis is devoted to the production of one or a few proteins and the identification and purification of mRNA is therefore made easier. In consequence, much of our knowledge of plant gene structure comes from the study of a relatively small number of proteins. These include: seed storage proteins from legumes, such as beans and peas, and cereals, such as maize (Messing

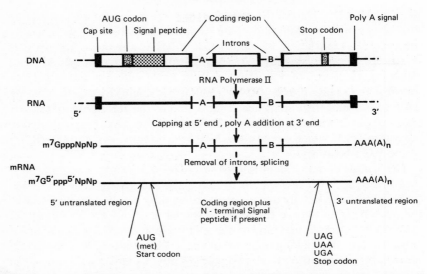

Figure 3.4 The relationship between the structure of plant nuclear genes and the structure of the corresponding mRNAs. Steps in the processing of mRNA are indicated. Note that not all plant genes coding for proteins contain introns (Figure 3.5).

et al., 1983; Larkins, 1983); leghaemoglobin, which is involved in nitrogen-fixing root nodules (Chapter 7); the small subunit of rubisco (Cashmore, 1983 and Chapter 4); seed lectin (Vodkin *et al.*, 1983); plant actin (Shah *et al.*, 1982) and a few others. The production of cDNA clones for these mRNAs (see Chapter 1) has led to the identification of corresponding genomic clones and the sequencing of these DNAs.

The comparison of many DNA sequences with the amino acid sequences of the corresponding proteins indicates that the plant nucleus–cytosol system uses the 'universal' genetic code. Some genes, such as the one for soybean actin (Figure 3.5), contain intervening sequences (introns) which interrupt the amino acid coding regions, whereas other genes such as those encoding the zein storage proteins (Figure 3.5) lack introns. When introns are present, they all contain the bases GU and AG at the left-hand and right-hand borders. The complete DNA sequence is transcribed as part of the mRNA precursor and the introns are subsequently cut out by processing enzymes in the nucleus and the exons (coding sequences) ligated together (see Figure 3.4). Recognition and excision of the introns has to be very precise; in some instances they actually split individual amino acid codons (Figure 3.5).

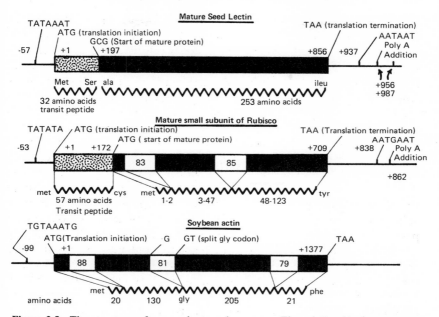

Figure 3.5 The structure of some plant nuclear genes. The relationship between gene structure and amino acid sequence of the proteins is shown for seed lectin (soybean), rubisco small subunit (pea) and actin (soybean). Nucleotide positions are shown in relation to the A of the ATG (met) initiation codon. Intervening sequences are shown white, with the number of nucleotides in each. Note the split gly codon between introns II and III of the actin gene. The 5′ end of the lectin mRNA has been mapped approximately 30 bases upstream from the ATG at the start of the transit peptide. There are no introns in the lectin gene. For further information about transcription initiation and polyadenylation sequences see Figure 3.6.

Some genes contain a nucleotide sequence coding for approximately 18–60 extra amino acids at the N-terminus of a protein which are not present in the mature protein. The extra amino acids function in a transient manner as a signal peptide, determining the transport of proteins into endoplasmic reticulum vesicles, as for the zein storage proteins, or as a transit peptide for transporting the protein across organelle membranes such as the chloroplast envelope (as for the small subunit of rubisco, Figure 3.5 and Chapter 4).

The signal peptides for the zein storage proteins are from 18–21 amino acids long and show slight variations in different genes (Messing *et al.*, 1983). The signal peptide for soybean actin, on the other hand, is 32 amino acids long (Shah *et al.*, 1982). During translation of mRNAs on the 80S ribosome the signal peptide is synthesized first and is recognized by the endoplasmic reticulum membrane. This recognition causes the nascent protein to bind to the membrane while still attached to the ribosome. The signal peptide is responsible for the transport of the protein through the membrane, into the lumen of the vesicle. During this process the signal peptide is removed by a membrane-associated protease. Some proteins which are transported across endoplasmic reticulum membranes become glycosylated in addition to having their signal peptides removed. This co-translational transport across the endoplasmic reticulum membrane contrasts with the post-translational mechanism of transport of proteins into chloroplasts. At least 100 nuclear-encoded chloroplast proteins are translated in the cytosol with transit peptides for crossing the chloroplast double membrane (Chapter 4). Recognition and transport do not require protein synthesis and removal of the transit peptide is carried out by a soluble protease in the chloroplast stroma. Comparison of the transit peptides from a number of chloroplast proteins suggest that they are highly variable in length and amino acid sequence and may have particular features related to the structure of the individual proteins. It is probable that there are may different types of signal or transit peptides for uptake of proteins into chloroplasts, mitochondria, glyoxisomes, peroxisomes, endoplasmic reticulum and secretion across the plasmalemma.

Many protein-coding sequences are present as a multigene family varying in size from two or three up to 100 related but not identical genes (Messing et al., 1983). This explains, for example, the variation in sequence of related storage proteins and the occurrence of isoenzymes. It presumably arises from gene duplication and divergence by mutation. Incomplete, and therefore inactive, genes called pseudogenes may also be present in the genome, alongside related functional genes (e.g. leghaemoglobin, Chapter 7). In addition, some normal genes may be inactivated by the presence of an insertion element (jumping gene) such as the transposable elements in maize (Chapter 2). Similar types of insertion elements are now being found by DNA sequencing studies in plants where there is no genetic evidence for jumping genes (Shah et al., 1982) and the phenomenon may turn out to be widespread. In situations where a gene contains an insertion element the plant may be unable to make the corresponding protein unless it carries a second normal gene.

3.4 Putative regulatory sequences in plant genes

A number of relatively short nucleotide sequences have been identified that are believed to be important for transcription, mRNA processing and translation. Variations on basic consensus sequences are found in all animal and plant genes (Figure 3.6). One of the putative transcriptional controls is the 'TATA box' which determines the site of transcription initiation. In plants this sequence is generally located 16–54 nucleotides upstream from the 'cap site' (5′ end of the mRNA). The actual sequence varies slightly in different plant genes and is related to a similar sequence in animals (Figure 3.6). The remaining 5′ (upstream) sequences are frequently AT-rich. There is evidence that the TATA sequence functions as a recognition site for RNA polymerase II in plants. Deletion of nucleotides in this region of the nopaline synthase gene of Ti plasmids (see Chapter 8) reduces or abolishes expression of the gene in plant cells. Furthermore, one of the maize zein genes has two TATA boxes, one 900 base pairs upstream from the first. Two types of mRNA (900 and 1800 bases long) are transcribed from this gene. The transcription start sites for the mRNAs

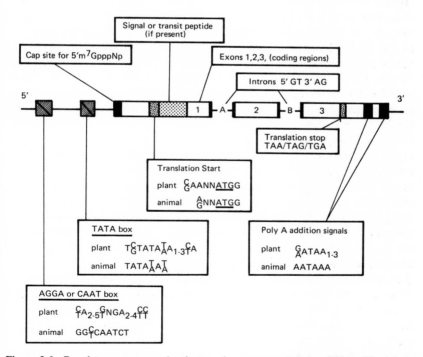

Figure 3.6 Regulatory sequences in plant nuclear genes and their mRNAs. The relative locations and consensus sequences for transcription initiation, mRNA processing and translation are shown and, where appropriate, animal and plant sequences are compared. Where variations in the consensus sequence occur at specific positions, alternatives are indicated. The numbers following particular bases show the extent to which that base may be repeated.

have been located by hybridizing them to the DNA and mapping the 5′ ends with S1 nuclease. In each case, the mRNAs start a short distance to the 3′ side of the TATA box (Langridge and Feix, 1983). A second possible transcription regulation sequence, the 'AGGA' or 'CAAT' box is located some distance upstream from the TATA box. Evidence that it is important comes from studies on animal gene expression which indicate that this region and other 'enhancer' sequences may regulate the extent of transcription. Although the 5′ end of mRNAs (the cap sites) can be located by S1 nuclease mapping, this is not necessarily the actual transcription start site, which consists of a 5′ nucleotide triphosphate. This can either be capped immediately or some nucleotides may be removed before capping of the remaining mRNA sequence occurs.

All plant mRNA genes examined so far have one or two poly-A addition signals in the 3′ untranslated region (Figure 3.6). The exact position of these signals relative to the protein synthesis termination codon and the poly-A tail varies a lot for different mRNAs and the actual sequence (5′ AATAAA 3′) may also vary slightly. Messenger RNAs can be terminated and polyadenylated some nucleotides after either the first or the second poly-A signal (Messing et al., 1983). A similar poly-A addition signal is found in the 3′ untranslated region of polyadenylated mRNAs from animals but the signal is not universal since it is absent from yeast mRNAs which contain poly-A sequences.

The base sequences of introns also vary a lot but they frequently consist of AT-rich DNA and contain a number of protein synthesis termination codons. The borders of plant introns always end in 5′ GT (GU in RNA) and 3′ AG and these nucleotides are generally conserved in all eukaryotes studied so far. This suggests there is a common mechanism for their recognition and excision. This may involve catalysis of the excision and ligation reaction by the intron border nucleotides themselves, as shown for the Tetrahymena rRNA introns (see section 3.2), or small nuclear RNA molecules which are complementary in sequence to the splice junctions. The processing reactions might occur by RNA-catalysed cleavage and ligation alone but probably also involve enzymes. It has been suggested (Sänger, 1982) that the pathogenicity of plant viroids (see Chapter 9) may be due to the fact that they interfere with mRNA splicing. Evidence for this view comes from the fact that they show sequence homology with small animal nuclear RNAs, such as U1, thought to be important for splicing.

In some cases (e.g. leghaemoglobin, Chapter 7), the introns in a gene are inserted between DNA sequences which code for separate domains of a protein and these are related to the protein structure. However, this is not always the case: there is no obvious relationship between the distribution of introns and the functional domains of actin genes. Furthermore, the zein storage proteins, which do contain different domains (Larkins, 1983) have no introns in the genes (Figure 3.5). It is possible that the occurrence and position of introns may reflect the way in which protein genes have evolved, perhaps by the juxtaposition of comparatively long DNA sequences coding for different peptides, thus producing more complex proteins. Introns can also play a more immediate role in metabolism, however. The first intron in the gene coding for subunit I of the yeast cytochrome oxidase actually codes for a 'maturase' protein which is involved in maturation or processing of the mRNA precursor. Mutations in this intron lead to failure to produce functional cytochrome oxidase subunit I (Chapter 5). Evidence for another intriguing aspect of the intron-exon relationship comes from studies with animal cells which show that different exons from one gene can be spliced together to make different versions of mRNA which code for related proteins with slightly different properties.

One feature of interest about the amino acid coding sequence is that the initiation codon, which is always for methionine, is surrounded by a particular pattern of bases which may be involved in recognition of the translation start site (Figure 3.6). The codon for termination of protein synthesis occurs some bases upstream from the poly-A signal and frequently there are multiple termination codons in this region.

3.5 Gene expression and plant development

Plant development frequently involves changes in gene expression which are triggered by environmental signals and alterations in the concentrations of endogenous hormones. There is now good evidence that this involves changes in gene transcription (Chapter 6) but we do not know how these are brought about. Control of gene expression may involve the reversible repression and activation of genes by methylation and de-methylation or changes in the state of the condensation of specific regions of the chromatin (Chapter 2). There may also be long-range interactions between widely separated DNA sequences which modulate transcription. It is also possible that there are specific promotor sequences that ensure that certain classes of genes are switched on at particular times. In addition there may be changes in regulatory proteins which alter the affinity of RNA polymerases for particular DNA sequences. In order to test these possibilities it is essential to examine a much wider range of plant genes than hitherto considered and to make use of developmental mutants in studies on gene expression.

CHAPTER FOUR

THE PLASTOME AND CHLOROPLAST BIOGENESIS

4.1 Plastid interrelationships

All meristematic plant cells contain small, relatively undifferentiated, organelles called proplastids which are capable of developing in a variety of ways. As the cells derived from the meristem differentiate to produce distinct tissues and organs, the proplastids also differentiate to form characteristic organelles such as amyloplasts, chloroplasts and chromoplasts (Table 4.1). Ultrastructural studies have shown that one type of plastid can develop into another and it is generally accepted that the various forms represent alternative states of differentiation of the same organelle. Most of our knowledge of the molecular biology of plastids has come from the study of chloroplasts and this is reflected in the attention given to these organelles in the succeeding pages.

Table 4.1 Types of plastid

	Location	Function
Proplastid	Meristematic cells	Progenitor of other plastids
Amyloplast	Cells near the growing tips of roots and shoots and starch-storing organs	The organelles contain starch grains and function as sedimentable statoliths for gravi-perception or in starch storage
Etioplast	Leaves of plants grown in the dark	Partially differentiated plastids converted to chloroplasts by light
Chloroplast	Leaves, stems and green fruits grown in the light	Photosynthesis, synthesis of lipids, amino acids, hormones, etc.
Chromoplast	Fruits and flower petals	Attraction of predators and pollinators

4.2 Chloroplast organization and function

Higher plant chloroplasts are approximately $5-10\,\mu$m long and are most frequently lens-shaped in section. They are surrounded by a double membrane and consist of a soluble phase, the stroma, and an internal membrane system, the thylakoids. The stroma contains the enzymes for carbon dioxide fixation and metabolism whereas the apparatus for the light reactions of photosynthesis is associated with the thylakoids (Figures 4.1 and 4.2). Broadly speaking, two separate strategies for carbon dioxide fixation are found in higher plants and these different mechanisms are reflected in chloroplast structure and composition. Plants of the C3 type, such as oats and soybean, have chloroplasts which store starch and in which the thylakoids come into close contact at intervals to form grana. In the C3 pathway (the Calvin Cycle) the chloroplast enzyme rubisco (ribulose bisphosphate carboxylase-oxygenase, previously known as fraction I protein, carboxydismutase or ribulose 1,5-diphosphate carboxylase) catalyses the addition of carbon dioxide to ribulose bisphosphate to form two molecules of phosphoglyceric acid (a three-carbon compound, hence the name C3).

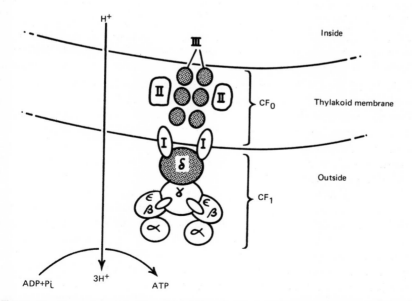

Figure 4.1 General organisation of the CF_0 and CF_1 coupling factors in the chloroplast thylakoid membranes. The subunits encoded by nuclear DNA are shaded.

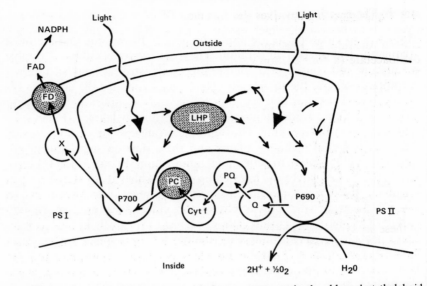

Figure 4.2 Trapping of light energy and electron transport in the chloroplast thylakoid membranes. Some of the proteins encoded by nuclear DNA are shaded: light harvesting chlorophyll a/b protein (LHP), plastocyanin (PC), ferridoxin (FD). The model illustrates the trapping of light energy by antenna chlorophylls, transmission to the P700 and P690 reaction centres of photosystems I and II (PSI and PSII), which is influenced by LHP, and the transport of electrons *via* various electron carriers, including plastoquinone (PQ) and cytochrome f (coded by the plastome).

Rubisco can account for up to 50% of the total soluble protein in a leaf and is probably the most abundant protein in nature (Ellis, 1981). It is a stromal protein with a molecular weight of over 500 000, consisting of eight large subunits (molecular weight 53 000) and eight small subunits (molecular weight 14 000). The active site is located on the large subunit. The small subunit is assumed to have a regulatory function. In addition to catalysing the carboxylation of ribulose diphosphate, the active site of rubisco can also function as an oxygenase. This reaction is important because it has the effect of reducing net carbon fixation in C3 plants by photorespiration. At low carbon dioxide concentrations or high oxygen concentrations, rubisco catalyses the oxidation of ribulose diphosphate to produce 3-phosphoglyceric acid and phosphoglycolic acid. The phosphoglycolate is subsequently metabolized in peroxisomes and mitochondria, resulting in the release of carbon dioxide. In C4 plants, such as maize and *Atriplex* species, the initial carbon dioxide fixation step occurs in leaf mesophyll cells containing chloroplasts which lack rubisco and do not store starch. The reaction involves the enzyme PEP carboxylase (phosphoenol pyruvate carboxylase), which catalyses the addition of one molecule of carbon dioxide to one molecule of phosphoenol pyruvate, to form oxaloacetate (a four-carbon compound, hence the name C4). The oxaloacetate is then converted to aspartic acid and malic acid which are transported to a ring of bundle sheath cells which surround the leaf vascular tissue, where they are decarboxylated. The resulting carbon dioxide is refixed by the bundle sheath chloroplasts which contain rubisco and operate the Calvin Cycle.

It is thought that the lack of photorespiration (and therefore greater efficiency) of C4 plants is due to the fact that decarboxylation of aspartic and malic acids maintains a high carbon dioxide concentration in the bundle sheath chloroplasts where rubisco is located.

There are several hundred individual polypeptides in chloroplasts. Some of these are synthesized by the organelles but others are encoded by nuclear genes and translated in the cytoplasm. It is a formidable task to work out their genetic origin, sites of synthesis, structure and function. They include more than 200 polypeptides in the stroma, involved mainly in carbon dioxide fixation and metabolism or in chloroplast protein synthesis. There are about 40 integral proteins in the two chloroplast envelope membranes, plus about 30 peripheral polypeptides. The envelope proteins include translocator proteins, which control permeability to small metabolites, plus enzymes involved in pigment synthesis. Finally, there are over 30 polypeptides present in thylakoids. Many of these are involved in the light reactions of photosynthesis, electron transport and ATP production (see Figures 4.1 and 4.2).

In addition to their role in photosynthesis and carbon metabolism, chloroplasts are involved in the synthesis of amino acids and nucleotides, proteins, nucleic acids, pigments, hormones, and in nitrate reduction. This necessitates integration of their metabolic activity with the other cell compartments. A more detailed account of chloroplast structure and biochemistry may be found in Reid and Leech (1980).

4.3 Chloroplast genetics and the extent of plastid autonomy

In the early part of the present century, when the full significance of Mendel's analysis of the inheritance of plant characters was appreciated, modern genetic studies began in earnest. Most nuclear genes follow the familar rules of Mendelian inheritance and, with the exception of sex-linkage phenomena in some organisms, similar results are obtained in crosses where the genotypes of the male and female parents are reversed. The inheritance of certain characters however, including some concerned with chloroplast development, is found to follow a different pattern, known variously as cytoplasmic, uniparental or maternal inheritance, where the genes for a particular character are only transmitted through the female parent. The explanation for cytoplasmic inheritance is that the female parent contributes most of the cytoplasm plus a haploid nucleus, whereas the male parent contributes mainly nuclear material; consequently, genes in the plastids and mitochondria are only passed on through the female line. It is now thought that in certain cases where the male parent does contribute significant cytoplasm, the accompanying organelle genomes may be destroyed, thus preserving the rule of maternal inheritance.

Maternal inheritance via chloroplast DNA (the plastome) has been demonstrated for some genes coding for chloroplast proteins (Table 4.2); other chloroplast genes control characters such as leaf size, shape and development and the production of variegations and chimeras. This evidence, together with the biochemical studies described in section 4.4 shows that the plastids carry genes which control events both inside and outside these organelles. Chloroplasts also contain the complete biochemical apparatus required for expression of the plastome, including DNA and RNA polymerases, ribosomes, tRNAs and tRNA-synthetases. There is a substantial body of evidence that they have evolved from captured free-

Table 4.2 Some of the chloroplast functions coded by the plastome

---

tRNAs

$\left.\begin{array}{l} \text{23S} \\ \text{16S} \\ \text{5S} \\ \text{4.5S} \end{array}\right\}$ rRNAs

Rubisco large subunit but not the small subunit

Cytochrome f

Cytochrome b_{559}

Elongation factors Ef-G, EF-T

P-700-chlorophyll a-proteins

'Photogene 32'

Proton-translocating ATPase (see Figure 4.1)

CF_o: subunits α, β, γ, ε but not δ subunit

CF_1: subunits I and III (the DCCD binding protein) but not subunit II

---

living photosynthetic prokaryotes. However, there are several arguments supporting the conclusion that the present-day plastids are not autonomous organelles but rely upon the nuclear genome for the determination of some of their functions. Firstly, the size of the plastome is insufficient to code for all the known chloroplast-located nucleic acids and proteins. Secondly, many genes determining plastid development and function have been shown to be inherited in a Mendelian fashion, which indicates that they are in the nucleus (Table 4.3). Thirdly, studies with mutant plants deficient in chloroplast ribosomes, and therefore unable to synthesize plastome coded proteins, nevertheless accumulate many chloroplast proteins. Finally, there is clear evidence from studies *in vitro* that many nuclear-coded proteins are translated outside the chloroplasts and are subsequently transported across the chloroplast envelope (see section 4.5).

One intriguing aspect of the cooperation of the nuclear genome and plastome in chloroplast biogenesis is that the two genetic centres sometimes code for different polypeptides which form part of the same enzyme. For example, in 1972 Kawashima and Wildman published evidence demonstrating that in tobacco the genes determining the primary structure of the small subunit of rubisco were inherited in a Mendelian fashion, and were therefore located in the nucleus, but those for the large subunit exhibited maternal inheritance, and thus were present on the plastome. More recently, the sites of synthesis of these polypeptides have been demonstrated by experiments *in vitro* and the respective genes have been cloned, sequenced and their physical locations confirmed (see section 4.4 and Chapter 3).

Table 4.3 Aspects of plastid function controlled by nuclear genes

Character or pathway	Number of genes analysed
Chlorophyll synthesis	14
Carotenoids in chloroplasts	5
Carotenoids in chromoplasts	6
Photosynthetic electron transport	15
CO_2 fixation and metabolism	5
Starch formation in amyloplasts	9
Chlorophyll-protein complexes	—
Variegations and chimeras	—
70S ribosomal proteins	20
Temperature sensitivity of plastids	—
Amino-acyl tRNA synthetases	—

4.4 Structure and expression of the plastome

Chloroplast DNA

The presence of DNA in chloroplasts was first demonstrated by Ris and Plaut in 1962. These researchers observed fibrils in electron micrographs of chloroplasts which could be removed by treatment with deoxyribonuclease. More recently, multiple copies of the plastid genome have been visualized by light microscopy, using DNA-binding fluorescent dyes. The molecules of DNA are attached to chloroplast membranes but extend into the stroma where they form nucleoids.

In higher plants, chloroplast DNA exists as a double-stranded circular molecule. Unlike nuclear DNA, it does not contain 5-methyl cytosine and is not complexed with histones. In many species the DNA has a buoyant density of approximately $1.697\,\mathrm{g\,ml^{-1}}$ in caesium chloride, which corresponds to 37% G + C, but values range from 36–40% in different species. In favourable circumstances, where the buoyant density of nuclear and chloroplast DNA are sufficiently different from each other, the two can be separated by isopycnic centrifugation in caesium chloride. In early experiments, the proportion of chloroplast DNA in a leaf was estimated to be about 1% of the total. We now know that this figure is far too low and the true value, based on measurements of the reassociation rate of total leaf DNA in the presence of a trace of radioactive chloroplast DNA, is between 10 and 20% (Scott and Possingham, 1982). This is in sharp contrast to the situation in roots, where plastid DNA only accounts for about 1% of total cellular DNA.

The potential coding capacity of the plastome has been measured by DNA–DNA reassociation, electron microscopy and restriction enzyme analysis. Calculations based on the reassociation rate of purified chloroplast DNA estimate the molecular weight to be between 80 and 100 million. which corresponds to between 120 000 and 150 000 base pairs. Direct measurement of the contour length of intact, circular, chloroplast DNA molecules with the aid of the electron microscope gives a value of between $37\,\mu\mathrm{m}$ and $55\,\mu\mathrm{m}$ for different species. These measurements are equivalent to molecular weight values of between 76 and 114 million, or 115 000 to 170 000 base pairs. The variation in the estimates made by different researchers is not explained solely by experimental error but reflects genuine differences in plastome size between species of higher plants. Recent estimates of plastome size in a number of different species, based on restriction endonuclease analysis, are given in Table 4.4.

Table 4.4 Size of the plastome in different plants determined by restriction endonuclease analysis

Species	Plastome size in kilobase pairs
Algae	
Chlamydomonas reinhardii	195
Codium fragile	85
Higher plants	
Cucumis sativus (melon)	155
Narcissus pseudonarcissus (daffodil)	161
Nicotiana tabacum (tobacco)	130
Pisum sativum (pea)	124*
Spinacia oleracea (spinach)	145
Triticum aestivum (wheat)	135
Vicia faba (broad bean)	121*
Zea mays (maize)	139

*These plants have only one copy of the rRNA-coding region which is repeated in most plants

It is generally assumed, on the basis of the developmental relationships between different types of plastid, that within a species the DNA in each type of plastid is identical. In the few plants where chromoplast DNA has been investigated it does seem to be very similar to chloroplast DNA. Furthermore, comparison of the restriction enzyme fragments of DNA from the chloroplasts of the bundle sheath and mesophyll cells of C4 plants has shown no differences.

Estimates of the coding capacity of the plastome, based on reassociation kinetics and electron microscopy, are in agreement and restriction endonuclease analysis shows that the yield and number of restriction fragments is that which would be expected from a uniform population of molecules of the size indicated in Table 4.4. Taken together, these data indicate that the chloroplasts contain one type of chromosome and are polyploid. In very young leaves each etioplast or chloroplast may contain 200 or more copies of the plastome. The DNA is replicated in a semi-conservative fashion. In chloroplasts of maize and pea, DNA replication begins at two sites about 7000 base pairs apart and proceeds in both directions. Multiple copies of the genome may also be synthesized by a rolling-circle mechanism (Kolodner and Tewari, 1975). However, during rapid leaf expansion, when the chloroplasts divide, DNA synthesis does not keep pace with the process of plastid division and older chloroplasts may only contain 20–30 copies of the plastome.

A restriction endonuclease map for maize chloroplast DNA (139 000 base pairs) is given in Figure 4.3, together with the location of a number of coding sequences. Similar maps are available for a number of other plants. The most striking feature of the plastome map is the presence of a 22 000-base-pair inverted repeated sequence, containing the rRNA genes. Similar repeats have been found in most plastomes, which thus carry two copies of

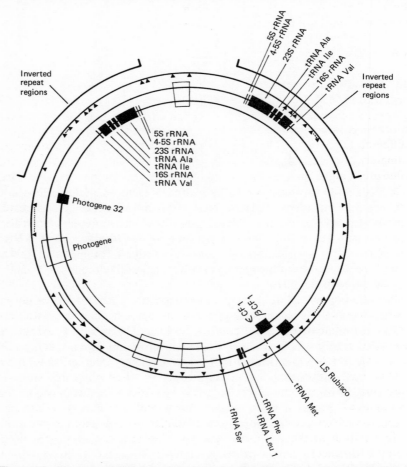

Figure 4.3 The maize plastid chromosome showing *Bam* H1 restriction enzyme sites (black triangles) and the location of some of the genes. Note the inverted repeat regions containing the rRNA genes, the location of some tRNA genes, rubisco large subunit, CF$_1$ polypeptides, photogene 32 and other unidentified photogenes (PG). The direction of transcription of the two DNA strands is shown by arrows. Where particular genes have not been assigned to one strand these are shown as boxes. (Modified from Bogorad *et al.*, 1983).

each rRNA gene. Notable exceptions are pea and broad bean, with only one copy, and some strains of *Euglena*, which have three copies plus an extra 16S rRNA gene, all with the same polarity. Although the plastome is a covalently closed circle, treatment with ribonuclease or alkali reveals the presence of ribonucleotides at between 12 and 18 sites. They are covalently linked to the DNA and occur on both strands. It is thought that these ribonucleotides are related to chloroplast DNA replication.

Chloroplast ribosomes and protein synthesis

In 1962 Lyttleton showed that chloroplasts contain ribosomes with a sedimentation value of 70S which can be separated from the larger 80S ribosomes of the surrounding cytosol in an ultracentrifuge. Structural and biochemical studies of the two types of ribosomes show that they are quite different (Table 4.5). The 70S chloroplast ribosomes measure approximately 20 nm across. They resemble more closely those of bacteria and blue-green algae than do the 80S ribosomes of the plant cytosol, and this similarity with prokaryotes extends to the general organization and sequence of the genes for chloroplast rRNAs which are located on the plastome (Figure 4.4). The sequences of the various rRNAs indicate that there is great potential for the formation of hairpin loops within each single-stranded molecule and thermal denaturation studies confirm that they have substantial secondary structure.

The chloroplast rRNAs are smaller than their counterparts in 80S ribosomes (Table 4.5) and the two types can readily be distinguished by gel electrophoresis. Roots contain very few 70S plastid ribosomes, whereas leaf cells contain several million, which account for between 30 and 50% of the total ribosome population of the cell (Figure 4.5). In some algae, chloroplast rRNA synthesis is strictly light-dependent and in most higher plants light stimulates the accumulation of chloroplast rRNA and ribosomes. However, substantial synthesis of plastid ribosomes also occurs in leaves of plants allowed to germinate and grow in darkness and etioplasts have the capacity to synthesize plastome-coded proteins.

In addition to the RNA components, chloroplast ribosomes contain about 50 distinct ribosomal proteins, distributed between the two subunits. The 23S, 5S and 4.5S rRNAs are present in the 50S subunit and the 16S rRNA is in the 30S subunit (Table 4.5). During protein synthesis, mRNA binds to the smaller ribosome subunit, assisted by hydrogen bonds between the 3′ end of the 16S rRNA and the 5′ untranslated region of the mRNA (section 4.4).

Table 4.5 Properties of chloroplast and cytosol ribosomes

	Source and sedimentation coefficient of ribosomes			
	Chloroplast 70S		Cytosol 80S	
	Large subunit 50S	*Small subunit 30S*	*Large subunit 60S*	*Small subunit 40S*
Initiating amino acid for protein synthesis	formyl methionine		methionine	
Antibiotics which specifically inhibit protein synthesis	D-threo chloramphenicol, lincomycin		cycloheximide	
Sedimentation coefficient of ribosome subunits	50S	30S	60S	40S
Sedimentation coefficient of ribosomal RNAs	23S 5S 4.5S	16S	25S 5.8S 5S	18S
Molecular weight of ribosomal RNA	1.05×10^6 3.94×10^4 $2.1\text{–}3.3 \times 10^4$	0.56×10^6	1.25×10^6 5.07×10^4 3.84×10^4	0.7×10^6
Approximate number of nucleotides (the number varies in different species)	2890 122 65–103	1490	3580 157 120	1926

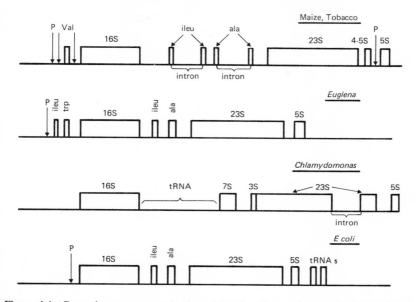

Figure 4.4 General sequence organization of rRNA coding regions in plastid DNA from different plants. Transcription promoter regions are indicated by P and an arrow. (Redrawn from Bohnert, 1983.)

In many higher plants the 23S chloroplast rRNA has a tendency to produce several large fragments. This is probably due to mild nuclease attack *in vivo* at RNA loops exposed near the ribosome surface. The individual fragments can be induced to remain together during RNA extraction and handling if precautions are taken to maintain the base-pairing between adjacent parts of the molecule. This is favoured by low temperatures and magnesium ions. At room temperature the short double-stranded regions melt and the fragments dissociate.

Plastids contain tRNA synthetase enzymes and unique tRNAs not found in the plant cytosol. Their structure (Dyer, 1982) and behaviour in heterologous charging experiments indicate that they are more closely related to counterparts in prokaryotic cells than those in the cytosol. Initiation of protein synthesis occurs with formylmethionine, as in bacteria, whereas methionine carries out this function in the cytosol. The presence in chloroplasts of tRNAs capable of being charged with all of the 20 protein amino acids has been demonstrated and isolated chloroplasts can carry out apparently normal protein synthesis. The use of codons for specifying amino acid incorporation during protein synthesis has been

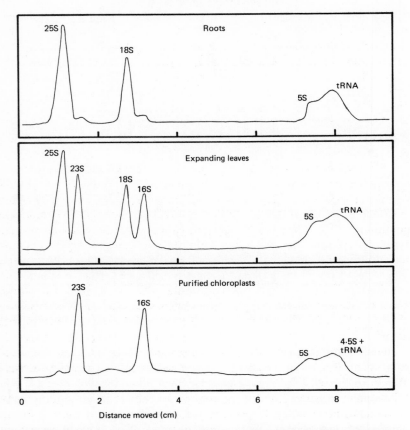

Figure 4.5 Comparison of the amount of plastid and cytosol rRNA in different plant tissues. RNA was extracted from roots and leaves of mung bean and fractionated by polyacrylamide gel electrophoresis together with RNA from a purified chloroplast fraction. Leaves of seedlings grown in the dark also contain substantial amounts of 23S and 16S plastid rRNA.

calculated from DNA sequences of plastid genes from spinach and maize. Although there is some bias in favour of A or U in the third base it appears as if chloroplasts, unlike mitochondria, use the normal genetic code (Bohnert *et al.*, 1982) (see Chapter 5).

Studies involving the inhibition of protein synthesis *in vivo*, employing antibiotics affecting either 70S or 80S ribosomes (Table 4.5), have been carried out in order to gain information about the sites of synthesis of chloroplast proteins. In one of the first studies of this type, Criddle *et al.*

(1970) showed that in barley the synthesis of the small subunit of rubisco is specifically inhibited by cycloheximide whereas the synthesis of the large subunit is inhibited only by chloramphenicol. This indicates that the mRNA for the large subunit is translated on 70S ribosomes inside the chloroplasts and the small subunit is translated by 80S ribosomes in the cytosol. Similar studies with inhibitors have been carried out for over 30 chloroplast proteins (summarized by Bottomley and Bohnert, 1982). However, conflicting results have sometimes been obtained and the general inhibitor approach has now been superseded by *in-vitro* methods which give unequivocal results.

Chloroplast protein synthesis has been studied *in vitro* either by resuspending the isolated organelles in a medium without osmoticum, thus disrupting the envelopes and making them permeable to externally added ATP and other substrates, or by placing intact organelles in an osmoticum containing KCl, when they retain the capacity for light-driven protein synthesis. Without osmoticum and light, the damaged chloroplasts require the addition of ATP to fuel protein synthesis. However, intact chloroplasts are capable of taking up added amino acids from the medium and incorporating these into proteins, using ATP generated by photo-phosphorylation (Blair and Ellis, 1973; Bottomley *et al.*, 1974). This light-driven protein synthesis provides a very useful control to ensure that incorporation measured is actually taking place in the chloroplasts and not in contaminating organelles or micro-organisms. It occurs in the absence of chloroplast RNA synthesis and is therefore directed by mRNAs already present in the chloroplasts before they are isolated.

Although it seems likely that polypeptides labelled during light-driven protein synthesis are synthesized on chloroplast ribosomes, it remains possible that mRNAs could cross the chloroplast envelope *in vivo*. Therefore, further proof of the location of the genes coding for such proteins is required before it can be concluded that they are coded for by chloroplast DNA. Proteins synthesized by isolated chloroplasts are generally studied by one- or two-dimensional electrophoresis in polyacryl-amide gels and identified by fingerprinting or by immunological pro-cedures using specific antibodies to known proteins.

It was initially thought that chloroplasts were capable of synthesizing only a few major polypeptides but further examination with more sensitive techniques showed that up to 80 radioactive polypeptides can be recovered after light-driven protein synthesis. Many of these are as yet unidentified and there is some doubt as to whether they are all authentic proteins or whether some of them might be artefacts of the methods used. This is an important issue since if all the proteins are genuine they could account for most of the coding capacity of the plastome.

The first polypeptide among the products of chloroplast protein synthesis *in vitro* to be characterized was the large subunit of rubisco from pea chloroplasts, which was identified on the basis of a tryptic peptide map by Blair and Ellis (1973). These researchers also showed that the small subunit of rubisco was not synthesized by isolated chloroplasts, which coincides with the conclusions from breeding experiments and the use of protein synthesis inhibitors. Similar results have been found subsequently with many plants. Notable exceptions, however, are the mesophyll cell chloroplasts of C4 plants, which do not synthesize the rubisco large subunit, although they contain the gene.

The small subunit of rubisco, together with many other chloroplast proteins, is coded for by a nuclear gene and translated in the cytosol on 80S ribosomes. The translation product contains a peptide sequence at the amino-terminus which specifies transport of the protein across the chloroplast envelope (Figure 4.10 and section 4.5). A most significant exception to the location of the rubisco small subunit gene in the nucleus has been found in the biflagellate alga *Cyanophora paradoxa*, which contains photosynthetic cyanelles which are thought to be endosymbiotic cyanobacteria. In this alga the cyanelle genome contains both the large and small subunits of rubisco (Heinhorst and Shively, 1983). It can be argued that this finding strengthens the speculation that plastid-encoded genes, originating from captured photosynthetic prokaryotes, have moved to the nucleus during plant evolution.

Other polypeptides that have been identified among the products of chloroplast protein synthesis (reviewed by Bottomley and Bohnert, 1982) are cytochrome f, cytochrome b_{559}, two elongation factors involved in chloroplast protein synthesis, one or two polypeptides of the P700-chlorophyll-protein complex, and several other membrane proteins including 'photogene 32' and components of the ATPase complex, discussed below.

The ATPase is involved in proton translocation across the thylakoid membrane, which is linked to ATP synthesis. It is composed of two multi-subunit coupling factors: CF_0, which is buried in the membrane, and CF_1, which is exposed (Figure 4.1). The alpha, beta, gamma and possibly the epsilon polypeptides of CF_1 and subunits I and III (called the DCCD binding protein) of CF_0 are synthesised in chloroplasts (Nelson *et al.*, 1980; Doherty and Gray, 1980). The delta subunit of CF_1 and polypeptide II of CF_0 are thought to be made in the cytosol. This contrasts with the situation found for the mitochondrial ATPase in yeast where all five CF_1 subunits are made in the cytosol and four of the CF_0 subunits are synthesized in the mitochondria.

Developing etioplasts accumulate several mRNAs coding for new proteins when they are illuminated (Chapter 6). Some have been located on the plastome map by hybridization and have been called photogenes. One of these, photogene 32, has been detected in spinach, pea, duckweed and maize. It codes for a 32 000 molecular weight thylakoid protein which is synthesized initially as a slightly larger protein (calculated to be 39 950 from the DNA sequence but, paradoxically, measured as 34 000 in SDS-polyacrylamide gels) and is processed before or during assembly into the thylakoids. The mRNA for photogene 32 is absent from etioplasts and accumulates in the light. Synthesis of the photogene 32 protein (also called psbA) occurs at very high rates but it must turn over very rapidly because it does not accumulate in large quantities. This protein is particularly interesting because it is involved in electron transport in photosystem II and it binds atrazine herbicides (Steinbeck et al., 1981). It is thus an obvious target for engineering herbicide-resistance (Chapter 10).

Another method for studying the synthesis of chloroplast proteins in vitro is to use a coupled transcription-translation system. There are several variations on this procedure but basically they make use of the similarities between chloroplasts and bacteria and employ E. coli RNA polymerase to transcribe purified chloroplast DNA or cloned restriction enzyme fragments, followed by translation of the RNA synthesized, using an in vitro protein-synthesis system. The advantage of this approach is that, if the protein product can be identified, it enables a specific gene to be located on a particular restriction enzyme fragment. A number of chloroplast proteins have been mapped in this way, including rubisco large subunit and some of the CF_1 polypeptides.

Transcription and processing of RNA

The chloroplast rRNA genes were first located on the plastome by saturation hybridization of chloroplast DNA with radioactive rRNA (Scott and Smillie, 1967). More recently the genetic arrangement has been studied by hybridization of rRNA to restriction enzyme fragments of the inverted repeat region of the plastome and by DNA sequencing. Maps of the plastome rRNA genes from a number of plants are shown in Figure 4.4 and compared with that for E. coli. The rRNA sequences occupy about 5700 base pairs in Euglena, but are longer in other plastomes which contain introns in some of the genes. The general arrangement of the genes is similar to that in *E. coli*.

The DNA sequences of the maize and tobacco 16S rRNA genes are 1491 and 1486 nucleotides in length respectively. They show 96% sequence homology with each other and 75% homology with *E. coli* 16S rRNA, which is 1542 nucleotides long. Similarly, the *E. coli* 23S rRNA gene, which is 2904 nucleotides long, has 67% homology with the 23S rRNAs from maize (2898 nucleotides) and tobacco (2804 nucleotides) (Edwards and Kossel, 1981; Takaiwa and Sugiura, 1982). The chloroplast and *E. coli* sequences have runs of up to 53 consecutive bases in common and are assumed to have arisen from a common ancestor (Schwarz and Kossel, 1980; Tohdoh and Sugiura, 1982). There are three short insertions in the maize 23S gene compared to *E. coli*. Each insertion has an inverted and a tandem repeat sequence reminiscent of transposable elements. The distance between the end of the 16S and the start of the 23S rRNA gene is 440 base pairs in *E. coli*, 259 in *Euglena*, 2408 in maize and 2080 in tobacco. The much longer sequence in the higher plants is due to the presence of introns ranging from 707 to 949 base pairs in the regions coding for the anticodon loops of the tRNA-ileu and tRNA-ala genes (Figure 4.4). In *Chlamydomonas* the 23S rRNA gene is also interrupted by an intron about 870 base pairs long, close to the 3′ end. This intron is flanked by 5′ CGT direct repeats and also has direct repeats of 5′ CGTGA at or near each end.

The 16S, 23S and 4.5S rRNA sequences in chloroplasts, together with the tRNAs in the spacer region between the 16S and 23S genes, are transcribed as a polycistronic precursor RNA which is subsequently processed or modified to produce mature rRNAs and tRNAs. Labelling of plants *in vivo* with ^{32}P-phosphate or ^{3}H-uridine has shown the presence of precursor RNA in spinach and mung bean chloroplasts. In spinach the precursor rRNA, with a molecular weight of 2.7×10^6 (approximately 8000 nucleotides), has been shown to be a product of chloroplast DNA since it accumulates in isolated chloroplasts carrying out light-driven RNA synthesis. Competition hybridization experiments have shown that the chloroplast precursor rRNA contains one copy of each of the 16S, 23S and 4.5S rRNA sequences, but not the 5S sequence. Studies on the kinetics of labelling of the various RNAs indicate that the 5S rRNA arises as a separate RNA precursor in chloroplasts of at least some plants. This is consistent with the existence of a putative RNA polymerase promoter sequence between the 4.5S and 5S genes in maize (Figure 4.4).

Transcription of the rRNA genes by chloroplast RNA polymerase begins at a promoter region upstream from the mature 16S rRNA sequence and continues past the end of the 4.5S sequence. The processing pathway for the primary transcript has not been fully worked out. In some plants the precursor may exist separately for some time, while in others it is processed very rapidly, perhaps even before synthesis is complete, as in *E. coli*. Processing involves the generation of a number of RNA fragments by the action of endonucleases (processing enzymes) at specific points in the molecule. A combination of *in-vivo* and *in-vitro* experiments (reviewed by Grierson, 1982) has shown the existence of a number of intermediates, including immediate precursors to 23S and 16S rRNA, called p23S and p16S. In *E. coli*, and almost certainly in chloroplasts also, the transcribed regions flanking the 5' and 3' ends of the 16S rRNA sequence in the precursor molecule can form a long double-stranded stem, with the 16S rRNA sequence looped out (Figure 4.6). RNA processing enzymes recognise a number of sites in this base-paired region and cut the molecule. A similar arrangement occurs during the processing of the 23S and 5S rRNA sequences. At the same time that this type of processing is going on, specific bases are methylated and proteins are added, as part of the ribosome assembly process.

Another type of processing reaction involves the removal of introns and the ligation of RNA sequences. This must happen for the *Chlamydomonas* 23S intron, since it is present in both gene copies but is not represented in mature 23S rRNA. The *Chlamydomonas* rRNA is also unusual in another respect, since it contains 7S and 3S sequences not found in other

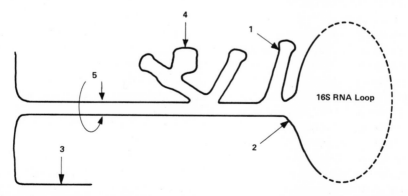

Figure 4.6 Probable secondary structure of part of the *E. coli* rRNA precursor showing base pairing of the regions flanking the 16S rRNA sequence. Site 1 shows the 5' end of the 16S rRNA, site 2 is the 3' end. Site 3 shows the 5' end of tRNA-ileu, Site 4 is the 5' end of p16S (precursor) RNA and Site 5 shows the site attacked by ribonuclease III. (Modified from Grierson, 1982.)

chloroplast ribosomes. The 4.5S RNA sequence, present in most chloroplast ribosomes, is of variable length in different plants (Table 4.5). It is an integral part of the p23S RNA and there is a short region of complementary between 4.5S RNA and 23S RNA. It is probable, therefore, that the 4.5S rRNA arises as part of a processing reaction in which intervening RNA is removed but the remaining segments are not ligated. A series of processing, excision and ligation reactions probably occurs for the interrupted tRNA genes in maize and tobacco. Many other tRNA genes are scattered around the plastome (Figure 4.3) and are presumably transcribed separately; by analogy with *E. coli*, some of these may also form parts of polycistronic transcription units. In at least some cases the terminal CCA of the tRNAs is not coded genetically and is presumably added after synthesis by a nucleotidyl transferase enzyme.

Not much is known about the synthesis and processing of chloroplast mRNAs *in vivo*. They seem to lack the 5′ cap ($m^7G^{5'}ppp^{5'}$) characteristic of cytosol mRNAs and do not contain long regions of polyadenylic acid at the 3′ end. However, there are several reports which suggest that chloroplast mRNA may contain short runs of oligo-A. Saturation hybridization experiments with chloroplast RNA suggest that from 70–90% of the DNA is transcribed, but it is not clear what proportion of this is mRNA.

DNA-dependent chloroplast RNA polymerases from a number of plants including maize and spinach have been solubilized and studied (reviewed by Wollgiehn, 1982). They are complex multisubunit enzymes and the function of the individual polypeptides is not fully understood. The maize chloroplast polymerase has a molecular weight of about 500 000 and contains many polypeptides ranging from 180 000 to 27 000, which are structurally unrelated to those from nuclear RNA polymerases. The spinach enzyme has seven subunits ranging from 155 000 to 26 000. The purified enzymes transcribe chloroplast rRNA genes preferentially *in vitro*; this presumably happens *in vivo* also, since about 90% of chloroplast RNA is ribosomal. Factors have been purified which cause the polymerase to transcribe certain genes preferentially. For example, an 'S factor' from maize chloroplasts, with a molecular weight of 27 000, causes the polymerase to favour the transcription of chloroplast sequences over other genes when both are present in chimaeric plasmids cloned in *E. coli*. This transcription specificity is greatest with supercoiled DNA, from which the chloroplast sequences are transcribed eight times more frequently than other sequences in the cloning vector (Bogorad *et al.*, 1983). Recent evidence suggests that there are differences between the solubility and specificity of the polymerases which transcribe chloroplast rRNA and tRNA sequences. However, it is not known whether this reflects differences in the core enzyme or in regulatory polypeptides associated with the polymerase. Other *in-vitro* transcription experiments have shown that the maize rubisco large subunit gene is transcribed three times more efficiently by chloroplast RNA polymerase in the presence of S factor than the beta and epsilon genes for the CF_1 complex (Bogorad *et al.*, 1983). It seems very probable that transcriptional regulation also occurs in chloroplasts *in vivo*. Hybridization with cloned chloroplast DNA probes shows that maize mesophyll cell chloroplasts do not contain rubisco large subunit mRNA whereas those in the bundle sheath cells do; furthermore, etioplasts contain no photogene 32 mRNA, which accumulates only in the light (see Chapter 6).

Putative regulatory sequences

Many plastome genes have now been sequenced, including those which code for identified products such as rRNAs, tRNAs, rubisco large subunit, CF_0 and CF_1 polypeptides, and photogene 32. A number of other

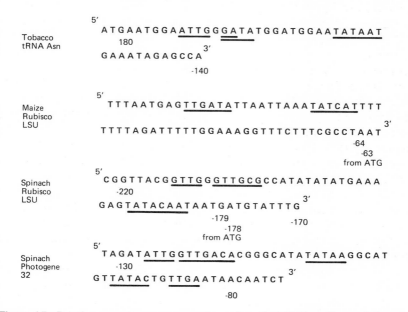

Figure 4.7 Putative promoter sequences upstream from the initiation of transcription of a number of chloroplast genes. (Modified from Bohnert *et al.*, 1982.)

sequences thought to represent as yet unidentified genes have also been determined. Collation of this information has led to the identification of possible control regions in DNA and in RNA transcripts deduced from the DNA sequences.

The rubisco large subunit gene from maize was the first to be sequenced (McIntosh *et al.*, 1980). Comparison of the coding region with the amino acid sequence of the protein shows that the normal genetic code is used and that there are no introns. Hybridization of rubisco large subunit mRNA to restriction fragments containing the gene enabled the 5′ end of the transcript to be mapped by S1 nuclease digestion. Transcription starts 63–64 nucleotides upstream from the ATG codon for the initiation of protein synthesis. Further upstream there are two putative promoter sequences (RNA polymerase binding sites, Figure 4.7) which are related to those in prokaryotes. Consensus sequences for these regions have been derived from the study of 15 maize genes (Bogorad *et al.*, 1983). Similar sequences are found in the regions upstream from the start of transcription of chloroplast genes from a number of different plants (Figure 4.7) but these are not at such a precise location with respect to the transcription start site as they are in *E. coli*. There is some evidence that these putative promoters actually function, since similar sequences which are upstream from the transcription initiation sites of chloroplast rRNA and tRNAs have been shown to be protected against deoxyribonuclease attack when RNA polymerases are bound to the DNA.

Sequences present in the 5′ untranslated region of the RNA transcript, close to the AUG codon for the initiation of protein synthesis, may function in recognition and binding of mRNA by the 3′ end of 16S rRNA (see

E coli 16S rRNA	5′ U A G G G G A A C C U G C G G U U G G A U C A C C U C C U U A 3′
Maize 16S rRNA chloroplast	5′ U A C U G G A A G G U G C G G C U G G A U C A C C U C C U U U 3′
Maize mRNA Rubisco large subunit	3′ G U A U U C A G G G A G G G A U G U U G A G U A C U U 5′
Spinach mRNA Rubisco large subunit	3′ G U A U U C A G G G A G G G A U G U U G A A G U U U A 5′

Figure 4.8 Comparison of the 3′ regions of *E. coli* and maize 16S rRNA with the 5′ non-translated regions of chloroplast mRNAs. Putative binding sites are underlined.

Figure 4.8) during protein synthesis. The GGAGG sequence in some chloroplast mRNAs is complementary to the CCUCC sequence in 16S chloroplast rRNA, but not all chloroplast mRNAs have these nucleotides at the 5' end. The CF_1 beta gene sequence, for example, has no GGAGG but does contain a TAGTG sequence (Bogorad *et al.*, 1983) which is complementary to a slightly different region of the 16S rRNA (Figure 4.8). It is possible that such differences in the 5' untranslated regions of mRNAs may affect their rate of translation.

Sequences in the untranslated region at the 3' end of some chloroplast genes have also provided interesting information. For example, there are two possible loop structures in the gene for maize rubisco large subunit in the region where the mRNA ends, approximately 98 nucleotides downstream from the termination codon for protein synthesis (Bogorad *et al.*, 1983). The same region also contains two types of repeated sequence. It is not known whether these features have significance for termination of transcription or for RNA processing but it is interesting that similar loops could occur at the 3' ends of other chloroplast mRNAs (Bohnert *et al.*, 1983).

A different situation is found at the end of the cistron coding for the beta subunit polypeptide of maize CF_1, which is actually fused to the start of the epsilon subunit. The ATG initiation codon for the second gene is formed by the last A of the lysine codon of the beta gene, plus the UG from the termination codon (Figure 4.9). Northern blotting experiments suggest there may be a dicistronic mRNA. It remains to be seen whether this arrangement is related to the coordination of synthesis of the two subunits. It is obviously not mandatory for functionally related genes to be fused however, since in pea chloroplast DNA the same genes are separated by 22

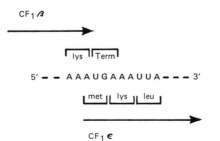

Figure 4.9 Fusion of the CF_1 β and ε genes in maize. The AUG initiator codon of ε is formed by the terminal A of the lysine plus the UG of the β gene termination codon. (Redrawn from Bogorad *et al.*, 1983.)

base pairs. Furthermore, in maize the cistron for the alpha subunit is located some distance away on the plastome.

The nucleotide sequence of the 5S rRNA from 70S ribosomes has been determined for a number of higher plants and shown to be substantially different from the sequence of the 5S rRNA of the 80S cytosol ribosomes (Dyer, 1982). The 5′ and 3′ ends of the molecule are thought to form a hydrogen-bonded stalk, with at least two other regions of base-pairing elsewhere in the molecule. The sequence CCGAAC, between residues 40 and 50, occurs in all prokaryotic 5S rRNAs, including those from chloroplasts. This sequence, which is not base-paired, is thought to interact with the GUψC sequence of tRNA molecules during protein synthesis.

4.5 Transit peptides and the genetic specification of protein transport into chloroplasts

Breeding experiments show (Table 4.3) that there are many nuclear genes which determine chloroplast functions and it follows that a mechanism exists for transporting macromolecules from the nucleus or the cytosol into chloroplasts. Present evidence indicates that this normally operates at the level of protein transport.

All nuclear-coded genes for chloroplast proteins investigated so far have mRNAs with a 5′ cap ($m^7G^{5'}ppp^{5'}$) and a 3′ poly-A sequence. They are translated on cytosol ribosomes as precursors containing transit peptides at the N-termini. This was first shown to be the case for rubisco small subunit (Dobberstein et al., 1977; Highfield and Ellis, 1978), which is initially translated as a 20 000 molecular weight polypeptide. In pea, the nuclear genes contain two introns, the first of which separates the transit peptide sequence from the region coding for the majority of the mature protein (Chapter 3). The transit peptide, which varies from 40–60 amino acids in different plants, is tailored to fit individual proteins, thus producing a particular configuration for transport. If amino acid analogues are incorporated into the sequence during in vitro protein synthesis, thus disturbing the normal secondary and tertiary structure, transport does not occur. The precursor protein crosses the chloroplast envelope double membrane by a post-translational mechanism, which requires ATP but not concomitant protein synthesis. Once inside the envelope, the transit peptide is removed by a soluble metallo-protease in the stroma. The processed rubisco small subunit is insoluble and is thought to bind to a 'small subunit binding protein' before combining with the large subunit and being assembled into the holoenzyme. A model for the synthesis and

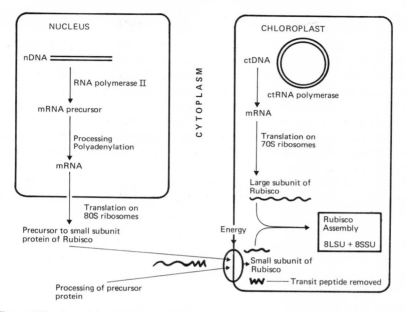

Figure 4.10 A model of the synthesis transport and assembly of ribulose bisphosphate carboxylase-oxygenase (rubisco). (Modified from Scott and Possingham, 1982.)

assembly of rubisco is shown in Figure 4.10. Almost all of the steps in this scheme have been shown to occur *in vitro*. Recently, the translation, uptake and processing of about 100 different chloroplast-located polypeptides has been demonstrated, including plastocyanin, ferridoxin, fructose 1,6 bis-phosphatase, two light-harvesting proteins and rubisco small subunit (Grossman *et al.*, 1982).

There is a profound imbalance between the number of plastome-coded and nuclear-coded genes in photosynthetic cells of higher plants. Since they contain on average 50 chloroplasts, each with 10–20 copies of the plastome, the chloroplast genes are repeated approximately 500–1000 times. In contrast, there are only a few copies of related, but not identical, genes for light-harvesting proteins and rubisco small subunit in each nucleus (Chapter 4). There is some evidence, from studies of plants with different levels of ploidy, that the number of nuclear genes may have a controlling influence on the production of rubisco, but there is not general agreement about this. Whatever the mechanism, it is clear that there must be some coordination of gene expression. Furthermore, light has an effect on mRNA levels for the light-harvesting proteins and rubisco small subunit (Chapter 6).

Although the present evidence indicates that the interaction between the nucleus and the plastome occurs through the transport of proteins and small molecules, the possible transmission and exchange of nucleic acid sequences cannot be excluded. It seems almost certain that transfer of genes from the plastome to the nucleus has occurred during the course of evolution. It is also possible that other types of information exchange can occur. The demonstration of the presence of chloroplast DNA sequences outside the plastids (see Ellis, 1982), the existence of plasmids in plant cell organelles (Chapter 5) and the demonstration of the existence of mobile genetic elements in nuclei (Chapter 2) raises a number of intriguing possibilities related to the transmission of genetic information between cell compartments.

CHAPTER FIVE

MITOCHONDRIAL DNA ORGANIZATION AND FUNCTION

Mitochondria contain their own DNA, transcription and translation apparatus and synthesize a relatively small number of polypeptides essential to their role in ATP production. The ribosomes are different from those in chloroplasts and the cytosol and a slightly different genetic code is used by mitochondria. Most of the mitochondrial proteins are encoded in nuclear DNA, translated in the cytosol on 80S ribosomes in a precursor form and transported into the organelles (Schatz and Butow, 1983). Although all types of mitochondria share some common features, plant mitochondrial DNA is very different from that in animals and lower organisms and the study of its organization and function is opening up whole new areas of plant research related to the control of development, susceptibility to disease, communication between organelles and the mobilization and transfer of genetic information.

5.1 Plant mitochondrial DNA

Size and composition

Mitochondrial DNA is generally circular and double-stranded. In most higher plants the DNA has a buoyant density of about $1.706 \, g \, ml^{-1}$ in CsCl, which corresponds to approximately 47% $G + C$. The amount of DNA in different organisms is exceptionally variable and ranges from about 16 kilobase pairs (kb) in humans to over 2000 kb in muskmelon (Table 5.1). Clearly the complexity of the mitochondrial DNA of higher plants is far greater than that in other organisms and it represents a considerable challenge to elucidate its function. Recent evidence indicates that there is one large circular 'master' chromosome and other smaller circles which each represent part of the large chromosome. In maize the main chromosome is

Table 5.1   Mitochondrial DNA from different organisms

	Number of base pairs	Number of different molecules per organelle
Higher plants		
Brassica sp.	218 000	3 (circular) 218kb, 135kb, 83kb
Maize	570 000 plus a variable number of plasmid-like DNAs from 1 400–6 000 bp	7 (circular) from 570kb–47kb up to 4 (circular or linear)
Muskmelon	2 400 000	?
Fungi		
Podospera anserina	juvenile 95 000 senescent 30 000 + 2 400	1 (circular) 2 (circular)
Saccharomyces cerevisiae	75 000	1 (circular)
Others		
Cow and man	16 000	1 (circular)

570 kb and contains five direct repeats and an inverted repeat sequence. There are six smaller circles which range from 47–503 kb. These are thought to be derived from the main circle by recombination events occurring between pairs of repeated elements. In *Brassica campestris* the main mitochondrial chromosome of 218 kb is circular and contains direct repeats of a 2000 kb sequence which separates regions of 135 kb and 83 kb (Palmer and Shields, 1984). The 83 kb and 135 kb regions also exist as separate circles and it is thought that they are derived from and can also recombine to form the larger master chromosome. The situation is less clear in some Cucurbitaceae where the amount of mitochondrial DNA can be much larger.

A number of functions have been assigned to plant mitochondrial DNA on the basis of *in vitro* protein synthesis, DNA–RNA hybridization and DNA sequencing, but these only account for a relatively small proportion of the DNA (Table 5.2). There are some repeated sequences but there do not appear to be the very long stretches of A + T-rich 'spacer' DNA found in mitochondria of *Saccharomyces cerevisiae*.

The first sequences to be located on the mitochondrial DNA were those encoding the RNA components of the 78S ribosomes which are different from the corresponding rRNAs in the cytosol and the chloroplasts (Table 4.5). The plant mitochondrial 26S and 18S rRNAs are larger than their counterparts in some other mitochondria and there is a unique 5S rRNA in higher plants (Table 5.2). In maize the 18S and 5S genes are closely linked but are separated from the gene for 26S rRNA. Sequence

Table 5.2   Genes located on plant mitochondrial DNA

one gene for each of the rRNAs:
  26S mol. wt. $1.12 - 1.16 \times 10^6$
  18S mol. wt. $0.69 - 0.78 \times 10^6$ $\Big\}$ closely linked
  5S mol. wt. 39,000

tRNA genes

genes for 20–30 hydrophobic membrane-associated proteins including:

	mol. wt.
α-subunit of the F_1-ATPase	58 000
Proteolipid proton channel of F_0 (DCCD binding protein)	8 000
apoprotein of cytochrome b	42 900
cytochrome c oxidase subunit I	38 000
subunit II	34 000
one ribosomal protein	

chloroplast genes (in maize)
  16S rRNA

  tRNA

  large subunit of rubisco

analysis of the 3′ end of the 18S rRNA gene and the 5S rRNA gene shows that they are both transcribed from the same DNA strand and are separated by 108 base pairs (Chao *et al.*, 1983). The 26S coding sequence is approximately 16 kilobase pairs away. The rRNA sequences have some homology with bacterial and chloroplast rRNA genes and the plant sequences are more like those in *E. coli* than are the mitochondrial sequences in fungi and animals. The 3′ end of the 18S rRNA does not have the CCUCC sequence found in chloroplast 16S rRNA which is thought to bind the 5′ end of chloroplast mRNAs (see Figure 4.8). However, there is a sequence 5′ UGAAT 3′ which may bind the 3′ ACTTA 5′ sequence found near the start of some plant mitochondrial mRNA sequences. The 108-base-pair region between the 18S and 5S rRNA genes contains a methionine codon (initiation for protein synthesis) followed by 34 other codons, but it is not known whether a polypeptide is produced from this region or indeed whether it has any function.

Promiscuous DNA

There is evidence that DNA sequences can move between organelles (called promiscuous DNA—Ellis, 1982). In maize for example, a 12 kb region homologous to chloroplast DNA has been shown to be inserted into unrelated sequences present in the mitochondrial genome (Stern and Lonsdale, 1982). Chloroplast gene sequences shown to be present in mitochondrial DNA by restriction mapping and hybridization include those for 16S rRNA, tRNA and the large subunit of rubisco (Chapter 4). There is also evidence for the presence of a nuclear gene in the mitochondria of *Neurospora*. It is not yet clear how this transfer of DNA has occurred. It could take place when organelles come in close contact, as they do from time to time, but it might also involve specific vehicles such as jumping genes (transposable elements) similar to those found in nuclei (Chapter 3) or plasmid-like DNAs. The latter are known to occur in plant mitochondria and there is evidence that they can associate and dissociate from the mitochondrial chromosome (see section 5.5). The mechanism of movement of promiscuous DNA raises some intriguing questions bearing on the exchange of genetic information during the course of evolution. It also has potentially important consequences for genetic engineering.

5.2 Mitochondrial protein synthesis *in vitro*

The coding capacity of plant mitochondrial DNA has been investigated by *in-vitro* protein synthesis. This involves the isolation of coupled mitochondria purified by gradient centrifugation and free of bacterial contamination. The incubation medium generally contains an osmoticum, various inorganic ions, a reducing agent, a respiratory substrate such as sodium succinate and amino acids. Energy for protein synthesis is provided by ATP generated by respiratory-chain-linked phosphorylation but ATP is also sometimes supplied externally.

Protein synthesis is monitored by measuring the incorporation of a radioactive amino acid such as ^{35}S-methionine. Contaminating bacteria can be detected by carrying out control experiments using acetate (a carbon source for bacteria but not a respiratory substrate) in place of succinate and the absence of cytosol protein synthesis can be confirmed by adding cycloheximide, a specific inhibitor of translation on 80S ribosomes. Analysis of the radioactive polypeptides synthesized by plant mitochondria by SDS-polyacrylamide gel electrophoresis shows that between 20 and 30 different proteins are made (Figure 5.2). The total number is not certain; some of those detected could be artefacts produced by premature termination, proteolytic cleavage or charge modification *in vitro*. It is clear however that plant mitochondria synthesize many more than the 8–10 polypeptides produced by animal and fungal mitochondria with smaller genomes. Due to the similarity between many mitochondrial polypeptides,

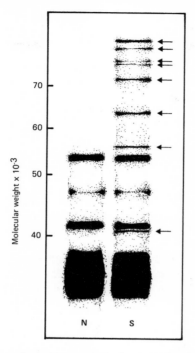

Figure 5.1 Radioactive polypeptides synthesized *in vitro* by mitochondria from maize shoots with normal (N) and male-sterile (S) cytoplasm. The polypeptides were fractionated by SDS-polyacrylamide gel electrophoresis and detected by autoradiography. (Redrawn from Forde and Leaver, 1979.)

it has proved possible to identify some of the proteins synthesized by plant mitochondria using cross-reacting antibodies and molecular probes developed during studies with other organisms. The proteins identified include: the DCCD binding protein of the proteolipid proton-channel of the F_0-ATPase (analogous to the DCCD-binding protein of the CF_0 complex of chloroplasts, Chapter 4); subunits I and II of the cytochrome c oxidase, the apoprotein of cytochrome b, and one ribosomal protein (Leaver *et al.*, 1983). In addition, plant mitochondria contain the gene for the α subunit of the F_1-ATPase (Hack and Leaver, 1983) and hybridization experiments suggest that a second gene coding for one of the other subunits $(\beta\gamma\delta\varepsilon)$ may also be present. In other organisms the F_1-ATP-ase genes are encoded in the nucleus and the location of some on plant mitochondrial

DNA raises interesting evolutionary questions. It also indicates that at least part of the extra mitochondrial DNA in plants is expressed.

5.3 Gene structure and mRNA processing

Several plant mitochondrial genes which code for proteins have been partially or completely sequenced including those for subunits I and II of cytochrome c oxidase and the apoprotein of cytochrome b (Fox and Leaver, 1981; Leaver *et al.*, 1983). These proteins show homology with their counterparts in mitochondria from other organisms. Comparisons of the amino acid sequences with the DNA sequences of the corresponding genes show that mitochondria do not use the 'universal' genetic code (Table 1.3) and that different organisms use various alternatives (Leaver *et al.*, 1983; Jukes, 1983). For example, in maize mitochondria CGG codes for tryptophan whereas in the universal code it represents arginine. AUG codes for methionine (the normal code), whereas in mammals and fungi AUA is used for methionine. In plant mitochondria UGG codes for tryptophan, which is the normal code, whereas fungi use what is normally a termination codon UGA to code for tryptophan. AUA codes for isoleucine (the normal code) whereas it codes for methionine in animals. So far no UGA termination codons have been found in plant mitochondrial genes; in animals and fungi UGA codes for tryptophan. It is also possible that mitochondria from different plants show variations, since *Oenothera* uses a CGG (normally arginine) in a position where a conserved tryptophan residue occurs in subunit II of cytochrome c oxidase.

The gene for the apoprotein of cytochrome b in maize is 1164 base pairs long and codes for a protein of molecular weight 42 900. The predicted amino acid sequence shows approximately 50% homology with the corresponding protein in yeast and cow. The gene has no introns in maize or cow but the related fungal genes do have introns. The major transcripts of the maize gene, revealed by Northern blotting, are 4.2 and 2.2 kilobases long, suggesting that there may be some processing of RNA transcripts. Similar studies of the gene for subunit II of cytochrome oxidase in maize indicates that the amino acid sequence has 47% and 40% homology with yeast and cow. In maize the gene contains a central 794-base pair intron but no introns are present in the corresponding genes from yeast, cow or another higher plant *Oenothera*. This raises the question of whether introns have any important function in transcription and processing. In maize the minimum length of a full transcript would be 1619 bases (including the intron) and 825 bases without the intron. Northern blotting has revealed transcripts of 3.2, 2.8, 2.45 and 1.95 kilobases. This also suggests the transcripts may be processed and that the mature mRNA contains long untranslated regions. However, it could also be that the initiation of transcription occurs at more than one site, as found for a nuclear storage protein gene (Chapter 3).

The sequence of the maize mitochondrial DNA coding for subunit I of cytochrome oxidase shows the gene is 1550 base pairs in length and contains no introns. Northern blotting experiments show that the most abundant transcript is 2.4 kilobases long. This contrasts with the situation in yeast mitochondria where the corresponding gene has between 7–9 introns (depending on the strain) and the first 4 or 6 contain long open reading frames (sequences beginning with a methionine codon and not followed by stop codons). The mutations have been characterized in yeast which result in failure to produce the cytochrome oxidase subunit I. All the mutations are located in the first intron, which codes for a 'maturase' or splicing enzyme involved in processing of the mRNA precursors (Carignari *et al.*, 1983). Thus the small amount of available evidence shows that the production of a specific protein does not *require* introns, but where they are present, their correct functioning *is* essential.

Very little is known about the structure or processing of plant mitochondrial mRNA. It is thought, however, that the 5′ end is not capped and that there is no extensive 3′ polyadenylation.

5.4 Plasmid-like DNAs in plant mitochondria

Senescence in Podospora anserina

In this fungus ageing is controlled by both cytoplasmic and nuclear genes. The mitochondria from juvenile mycelia contain 94 kilobase pairs of DNA (sometimes called the chondriome) whereas those from senescent mycelia have only 30 kilobase pairs. Mitochondria from sensecent mycelia also contain a small covalently closed circular (ccc) plasmid-like DNA of 2.4 kilobase pairs. This plasmid appears to be involved in senescence, since when it is introduced into juvenile fungal protoplasts by transformation it induces senescence. Restriction enzyme analysis and Southern blotting have shown that in juvenile mycelia the plasmid DNA is nevertheless present but integrated into the chondriome. It appears to become excised during senescence and can then exist freely and replicate (Kück *et al.*, 1981). During this process there is a reduction in size of the chondriome. It is not known whether a similar type of DNA rearrangement operates during the senescence of the cells of higher plants. However, higher plant mitochondria do contain apparently mobile genetic elements which seem to be important in determining male-sterility.

5.5 Mitochondrial DNA and cytoplasmic male sterility

In many plants the inability to produce fertile pollen is controlled genetically. It results from pollen abortion at one of several stages between meiosis and microspore mitosis and is governed by nuclear and cytoplasmic genes. In all the examples examined so far, cytoplasmic male-sterility is determined by genes present in the mitochondria and can be modified by nuclear restorer (Rf) genes. The phenomenon is made use of in the production of F_1 hybrids during the breeding of crop plants such as maize, sugar-beet, sunflower and *Sorghum*. In these plants the genes concerned are located in the mitochondria. (It was originally suggested that cytoplasmic male-sterility in tobacco is controlled by chloroplast DNA but this is now thought to be incorrect.)

Cytoplasmic male-sterility has been extensively studied in maize, where four general types of cytoplasm have been distinguished, called N, T, C and S. The normal (N) type gives rise to functional pollen whereas T, C and S plants are male-sterile and have cytoplasms which can be distinguished from each other on the basis of the mode of restoration of fertility by nuclear Rf genes. There are several lines of evidence linking male-sterility in maize with mitochondria. (1) Electron microscopy shows that the degeneration of mitochondria is one of the earliest events in pollen abortion. (2) Mitochondria from male-sterile plants synthesize different proteins *in vitro* compared to those from male-fertile plants. (3) There are differences in mitochondrial DNA between normal and male-sterile plants.

Analysis of DNA from maize mitochondria by agarose gel electrophoresis, without treatment with restriction endonucleases, shows that a variable number of low molecular weight plasmid-like DNA molecules are present (Table 5.3). Similar DNAs have been found in mitochondria of other plants including sugar-beet and *Sorghum*.

Mitochondria from maize plants carrying the S cytoplasm contain two linear DNAs called S-1 and S-2 (Table 5.3) which have a terminal 1500 bp sequence in common. These molecules, which can be 5–15% of the mitochondrial DNA, show some similarity to adenoviruses in that they contain a protein attached to the 5′ ends and have terminal inverted repeats of either 168 or 196 bp. Mitochondria containing free S-1 and S-2 DNAs synthesize eight additional proteins *in vitro* and seven of these are larger

Table 5.3 Occurrence of low molecular weight DNA molecules in maize mitochondria

DNA length in kilobase pairs	Configuration	Cytoplasm			
		N	T	C	S
6.2	linear	−	−	−	+ (S1)
5.2	linear	−	−	−	+ (S2)
2.35	—	+	−	+	+
1.94	supercoiled circle	+	+	+	+
1.57	circular	−	−	+	−
1.42	circular	−	−	+	−

than the normal mitochondrial proteins (Figure 5.1). It remains to be seen if the genes for these proteins are located on the S-1 and S-2 DNAs and whether their expression is related in any way to the male-sterile phenotype.

Some maize plants with S cytoplasm can revert to the fertile condition. When this happens the S-1 and S-2 sequences are no longer found separately. However, they are not totally lost but exist in a different state in the mitochondria. Restriction enzyme analysis shows that additional DNA sequences are present in the main mitochondrial DNA of such reverted plants and hybridization with labelled probes has enabled these to be identified as integrated S-1 and S-2 DNAs. The main mitochondrial DNAs of normal plants also contain sequences related to S-1 and S-2 at two separate sites. It has been suggested on the basis of these results that S-1 and S-2 may be capable of integration into and excision from the main mitochondrial DNA (Leaver and Gray, 1982).

Variations in maize mitochondrial protein synthesis are also found with organelles isolated from C and T male-sterile plants. Although these lack S-1 and S-2 molecules they do contain other low molecular weight DNAs (Table 5.3). In the case of T mitochondria, a 21 000 molecular weight protein which is synthesized by N mitochondria is absent and a new protein with a molecular weight of 13 000 is synthesized in its place. In addition, a 15 500 molecular weight polypeptide synthesized in N mitochondria is replaced in C mitochondria by a larger protein of 17 500 molecular weight (Figure 5.2). There is strong circumstantial evidence that the synthesis of these variant proteins is related to male sterility since they are characteristic features of all male-sterile plants of a particular group. Furthermore, the expression of the 13 000 molecular weight polypeptide in T mitochondria is reduced in organelles isolated from plants restored to fertility with nuclear Rf genes. However, in these restored plants the synthesis of the normal 21 000 molecular weight protein is not observed. The molecular basis of these changes is not fully understood. However, the demonstration that a

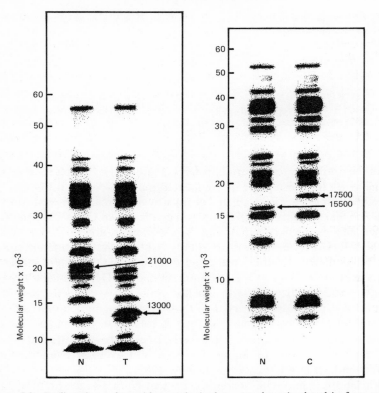

Figure 5.2 Radioactive polypeptides synthesized *in vitro* by mitochondria from maize shoots with normal (N) and male-sterile (T and C) cytoplasm. After electrophoresis in SDS-polyacrylamide gels the polypeptides were detected by autoradiography. (Redrawn from Forde and Leaver, 1979.)

nuclear gene has a specific biochemical effect in mitochondria opens up the possibility of studying nuclear control of plant organelle genome expression at the molecular level. This control could involve the transfer of nucleic acid to the mitochondria but most probably operates by the synthesis and transport of a protein into the organelles which is involved in regulation of mRNA transcription or processing. In a male-sterile *Sorghum* line it has been shown that the mitochondria synthesize a 42 000 molecular weight protein which resembles the 38 000 molecular weight subunit I of cytochrome oxidase (Table 5.2). The current view is that this change may result from altered processing of a mRNA precursor which gives rise to a larger protein. In some way, the alteration in the protein confers male sterility.

Maize plants carrying the T-type cytoplasm are susceptible to the fungus *Helminthosporium (Bipolaris) maydis* race T, which causes southern corn leaf blight. In 1970 approximately 80% of the maize grown in the United States carried the T cytoplasm and about 20% of the crop was lost, estimated at a value of one billion dollars. Severe symptoms on plants with T cytoplasm are caused by a toxin produced by the fungus. The toxin seems to affect the mitochondria since organelles isolated from T-type male-sterile plants show uncoupling of oxidative phosphorylation, swelling, disruption and a number of other effects when treated with the toxin. The presence of Rf genes, or genes closely linked to these in plants with T cytoplasm reduces the sensitivity of the plants to the fungus.

The sensitivity of T mitochondria to T toxin and the degeneration of mitochondria in the microspores after meiosis in healthy plants with T cytoplasm suggest that the two processes might be related in some way. It has been postulated that during pollen formation a compound is produced naturally in the anther which has the same effect on the mitochondria as the *H. maydis* T toxin.

Conclusions

One area for future research will be in defining further the role of the surprisingly large amount of DNA present in plant mitochondria. This promises to shed some light on the question of the evolution of mitochondria and the exchange of genetic information between organelles. In addition, further studies of the role of small DNAs in controlling processes such as male sterility and susceptibility to disease should pay dividends. There are obvious implications for the design of other methods of producing male sterility, for understanding sensitivity to disease and for genetic engineering.

CHAPTER SIX

HORMONES AND ENVIRONMENTAL FACTORS AFFECTING GENE EXPRESSION

Over 30 years ago it was demonstrated that cells removed from a differentiated plant and cultured under appropriate conditions are capable of regenerating to produce a complete new individual. This can now be achieved with single cells and protoplasts from a variety of plant species and tissues. It demonstrates clearly that many differentiated cells remain totipotent and it therefore follows that development and differentiation involve the selective control of gene expression.

Plant growth and development is regulated by five major hormones or groups of hormones: auxins, gibberellins, cytokinins, abscisic acid and ethylene. It is also influenced by environmental factors such as light quality, photoperiod, gravity and temperature. Some responses to the environment are governed by specific sensors and of these the red/far-red photo-reversible pigment phytochrome is probably the best understood. There is now indisputable evidence that these agents have important effects on gene expression. Unfortunately we do not know the *mechanism* of action of any of them. Molecular biology will not necessarily solve all the problems but it can provide suitable experimental approaches and furnish probes with which to dissect various aspects of development and hormone action. In this chapter we consider a number of examples which illustrate ways in which control of development is being studied.

6.1 Mobilization of reserves during germination of cereals

Seeds grown for food generally contain large quantities of stored carbo-hydrate, protein and other reserves. In cereals the main reserve material is starch, which accumulates in the amyloplasts of the endosperm cells during grain filling. The quality and quantity of starch produced are important factors in determining yield and the suitability of the cereal for certain types of food processing. The protein content is also important because of its nutritional value and specific properties it can confer on the product, for example during breadmaking or brewing. Storage proteins are synthesized by cereals during a particular stage in grain development and deposited in protein bodies, sometimes surrounded by a membrane derived from the endoplasmic reticulum or the tonoplast (Chapter 3). These food reserves stored in the endosperm are used by the growing embryo during germination. The process of mobilization of the reserves has been studied in a number of cereals, particularly in barley where it is important for malting.

When barley grains (Figure 6.1) imbibe water, metabolic activity quickly develops in the cells of the embryo and the aleurone layer (but not the endosperm, which at this stage is moribund). After a few days, mobilization of the stored reserves is quite noticeable and the endosperm becomes partly liquefied. Transverse sectioning of the endosperm at different times after imbibition shows that the solubilization process begins at the outside, in association with the cells of the aleurone layer, and progresses inwards. No solubilization is found if the embryo is removed from the endosperm at an early stage in imbibition. These observations led to the demonstration of a diffusible stimulus moving from the embryo to the aleurone layer which was identified chemically as gibberellic acid (GA_3). GA_3 is one of over 50

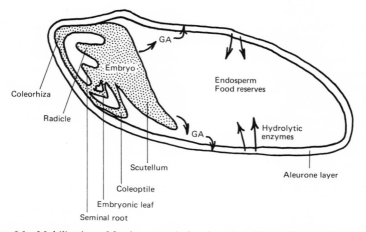

Figure 6.1 Mobilization of food reserves in barely grains. The synthesis and secretion of gibberellic acid by the scutellum of the barley embryo stimulates the aleurone layer to synthesize and secrete hydrolytic enzymes which digest the stored food reserves in the endosperm.

naturally-occurring gibberellins, many of which have important effects on plant growth and development. It is synthesized by the scutellum shortly after imbibition and its role in the mobilization of barley grain reserves has been studied using isolated aleurone layers.

Cells of the aleurone are small with thick walls and a densely-packed cytoplasm. In barley the aleurone layer is three cells thick and can be isolated from the rest of the grain. GA_3 has two separate effects on isolated aleurone layers: it causes the production of new hydrolytic enzymes and also brings about the establishment of mechanisms for the secretion and release of new and pre-existing enzymes to the outside of the cells. Incubation of aleurone layers in water leads to the production of a number of hydrolytic enzymes including acid phosphatase, β-1,3-glucanase and ribonuclease. Addition of GA_3 results in the release of these enzymes from the cells and also causes the production and release of α-amylase and protease. All of these hydrolytic enzymes are synthesized *de novo* during germination but only α-amylase and protease require the presence of gibberellic acid. One aspect of GA_3 action related to enzyme secretion is the substantial synthesis of endoplasmic reticulum elicited by the hormone. This is associated with increased membrane synthesis, formation of golgi and the stimulation of polyribosome formation. The synthesis of poly A-containing mRNA, including that coding for α-amylase, is also promoted by GA_3. Higgins *et al.* (1976) studied the appearance of translatable mRNA for α-amylase in response to GA_3 by immunoprecipitation of the proteins synthesized *in vitro* using a specific α-amylase antibody (Figure 6.2). More recently α-amylase mRNA has been measured by hybridizing nick-translated cDNA probes to aleurone layer RNA after Northern blotting. This more sensitive technique shows that α-amylase mRNA appears within one hour of applying GA_3 to aleurone layers and continues to increase with time. The appearance of the mRNA is prevented if cycloheximide is added at the same time as GA_3 (Muthukrishnan and Chandra, 1983). This suggests that there is a requirement for protein synthesis in order for α-amylase mRNA to be made. Such a protein might function as a receptor for GA_3, in promoting transcription of the α-amylase genes or in stabilizing the mRNA.

The size of the barley α-amylase mRNA translation product *in vitro* suggests it is synthesized as a precursor protein and therefore might contain a signal peptide as it does in wheat (Boston *et al.*, 1982). Although the enzyme is found associated with the endoplasmic reticulum *in vivo* (Jones and Jacobsen, 1982) the details of its synthesis and transport are not clear. There are known to be five isoenzymes which fall into two groups, based on

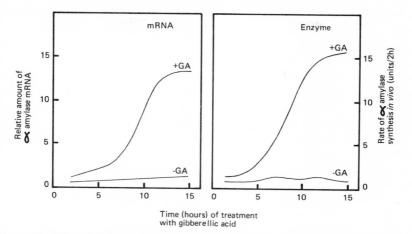

Figure 6.2 Effect of gibberellic acid on the accumulation of α-amylase mRNA and rate of α-amylase synthesis in isolated aleurone layers of barley. (Redrawn from Higgins *et al.*, 1976.)

immunological studies, and it appears that a small family of genes might be involved.

6.2 Effect of light on synthesis of chloroplast proteins

During seed germination and leaf growth in darkness the proplastids increase greatly in size and develop into etioplasts, which are several μm long. They contain large amounts of internal membrane including a characteristic prolamellar body formed by a paracrystalline lattice of tubules. Etioplasts frequently contain enzymes involved in CO_2 fixation and metabolism and some of the proteins normally found in the thylakoids of chloroplasts. With one or two exceptions they lack chlorophyll and some other components required for photosynthesis and only undergo a transformation to photosynthetic competence upon illumination. The responses of etioplasts to light are complex and several photoreceptors seem to be involved. These include the chlorophyll precursor pro-tochlorophyllide a and the photoconvertible pigment phytochrome, which produces a response after irradiation with red light (λ_{max} 660 nm) but not after far-red light (λ_{max} 730 nm).

The most obvious ultrastructural change following illumination is the dispersal of the prolamellar body and an increase in the amount of thylakoid membrane. Small amounts of chlorophyll a are formed rapidly by the light-stimulated reduction of protochlorophyllide *a* to chlorophyllide *a* by the thylakoid enzyme NADPH-protochlorophyllide oxidoreductase, which in most plants requires light in order to reduce its substrate. The resulting chlorophyllide *a* is then esterified with geranyl geraniol before reduction to produce the phytol side-chain of chlorophyll *a*. Photoactivation of phytochrome results in the subsequent stimulation of the chlorophyll biosynthetic pathway and chlorophyll-binding proteins also accumulate. Over a period of several hours more membrane material is synthesized, the thylakoids form grana and the familiar chloroplast ultrastructure and biochemical activity develops. During this process at least two thylakoid proteins decline in quantity and 20 or more appear. In addition, the synthesis of rubisco and other soluble Calvin Cycle enzymes is stimulated by light although they are generally present in small quantities in the dark.

Light affects the expression of genes both in the nucleus and plastids. At least five transcripts of plastome-coded photogenes appear during illumination (see Chapter 4). One of these, photogene 32, codes for the atrazine-sensitive 32 000 molecular weight thylakoid protein involved in photosystem II electron transport (Bogorad *et al.*, 1983). The mRNA for this protein accumulates in quite large quantities following illumination. It has been identified by hybridization to restriction enzyme fragments of plastid DNA in maize and by *in-vitro* translation of plastid mRNA from a number of plants. In duckweed, the rate of labelling of the mRNA *in vivo* has been shown to be correlated with the synthesis of the photogene 32 product (Rosner *et al.*, 1975; Edelman and Reisfeld, 1980) Translation *in vitro* shows that the protein is made in precursor form, about 8% larger than the mature protein. The precursor is presumably processed before or during insertion into the thylakoids. Although it is synthesized at very high rates, the protein turns over very rapidly and does not accumulate in large amounts.

Another important thylakoid protein synthesized in response to light is the light-harvesting chlorophyll *a/b* protein (LHP). Most plants contain two or three of these proteins, which are closely related, judged by amino acid composition, fingerprinting and immunology. The genes coding for the LHP(s) are inherited in a Mendelian fashion and the mRNAs have been shown to be synthesized by isolated nuclei. They contain poly-A and are translated in the cytosol. Little or no mRNA is present in the dark but it accumulates rapidly upon illumination, in response to photoactivation of phytochrome by red light (Figure 6.3 and Apel, 1979). The *in-vitro* translation products of the mRNAs have molecular weights of between

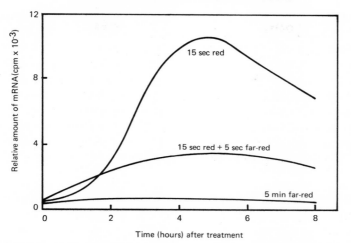

Figure 6.3 Effects of light on the accumulation of mRNA for the light-harvesting proteins in leaves of dark-grown barley seedlings. The mRNA fraction was purified, translated *in vitro* and LHP mRNA translation products precipitated with a specific antibody. The accumulation of mRNA is stimulated by red light and the effect is partly reversed by far-red light. (Redrawn from Apel, 1979.)

29 000 and 32 000, which is 4 000–6 000 greater than the mature proteins. The extra amino acids function as a transit peptide for transport of the protein across the plastid envelope (see Chapter 4). Although the mRNAs appear after illumination the light-harvesting proteins do not accumulate unless chlorophyll synthesis occurs; in the absence of chlorophyll the proteins are unstable and are degraded. Providing chlorophyll is present, the proteins are inserted into the thylakoids where they combine with chlorophylls *a* and *b*.

Between 10 and 20 of the N-terminal amino acids of the mature light-harvesting proteins are exposed at the membrane surface. This region of the LHP can be phosphorylated by a light-activated protein kinase and dephosphorylated by a phosphatase. Both enzymes are present in the chloroplasts and seem to respond to the redox state of plastoquinone. Reversible modification of the LHP is thought to control light energy distribution between photosystems I and II, thus maximizing the photosynthetic quantum yield (Bennett, 1983).

Both rubisco small subunit (Chapter 4) and the light-harvesting proteins are encoded by small families of nuclear genes. Although rubisco is frequently synthesized by dark grown plants and accumulates in etioplasts, its synthesis is enhanced by light, probably acting through the phytochrome system. In some plants, notably pea, the synthesis of rubisco seems to be light-dependent.

Cloned DNA probes have been used to measure the activity of nuclear genes coding for chloroplast proteins in isolated pea nuclei (Gallagher and Ellis, 1982). Transcription of the LHP genes is nine times higher and the rubisco small subunit genes 18 times higher in nuclei isolated from light-grown plants compared to those from plants grown in the dark. This strongly suggests a selective control of transcription after illumination since rRNA sequences are only transcribed twice as effectively in the light.

6.3 Ethylene and fruit ripening

Fruits are classified as 'climacteric' or 'non-climacteric', depending on whether or not they show a respiration increase (the respiratory climacteric) during ripening. In climacteric fruits such as apples, pears, bananas and tomatoes the first indication of ripening is an increase in carbon dioxide output and the synthesis of ethylene (Figure 6.4). Non-climacteric fruits, on the other hand, such as oranges, lemons, grapefruit and strawberries, show no increase during ripening. There is still some disagreement about whether respiration or ethylene production increases first in climacteric fruits. This is probably explained by the difficulties in measuring small changes in gas production by bulky organs. However recent experiments with tomatoes show that a large increase in natural ethylene synthesis precedes the respiratory climacteric. Furthermore, the application of ethylene to many plant tissues can *cause* an increase in respiration. The critical role played by ethylene in the stimulation of ripening is highlighted by the importance attached to controlling its synthesis and accumulation during fruit storage. The gas appears to act as a ripening hormone in climacteric fruits and ethylene has been used for many years for the commerical ripening of bananas.

One early view of ripening was that it is a degradative process brought about by a breakdown in 'organizational resistance' of the cell. This was thought to lead to leakage of various components across membranes and the activation or release of hydrolytic enzymes which rampaged through the cell catalysing autolysis. However, the existence of mutations which interfere with the ripening process makes it quite obvious that gene expression is also involved. For example, tomatoes normally take approximately seven weeks to grow and mature and then ripen over a period of 5–10 days. In contrast, fruits homozygous for the 'ripening inhibitor' (*rin*)

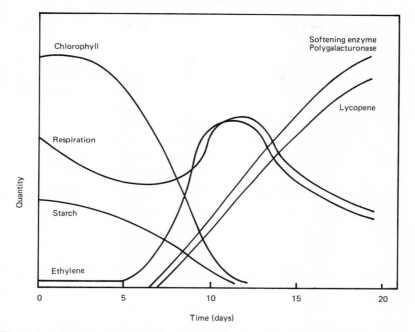

Figure 6.4 Some of the changes that occur during the ripening of tomato fruit.

mutation (on chromosome 5) make no lycopene and hardly soften at all. It is quite common for them to remain in good condition for many months. There is also biochemical evidence that normal ripening in tomatoes involves the continued synthesis of rRNA, tRNA, mRNA and that *de novo* enzyme synthesis is required to catalyse ripening changes (Grierson *et al.*, 1981; Grierson, 1984).

One important enzyme that is synthesized during tomato ripening is polygalacturonase (PG). The enzyme exists in three isoenzyme forms which are structurally and immunologically related. PG is absent from green fruit and is synthesized *de novo* during ripening. The 'neverripe' (*Nr*) mutant, which softens slowly, makes much less PG than normal and the *rin* mutant makes none (Figure 6.5). Purified PG isoenzymes have been shown to degrade the cell walls of mature green fruit *in vitro* and the role of the enzyme in degrading the pectin fraction of the cell wall during softening has been established. During normal ripening the synthesis of PG is preceded by ethylene production and treatment of mature unripe fruit with ethylene

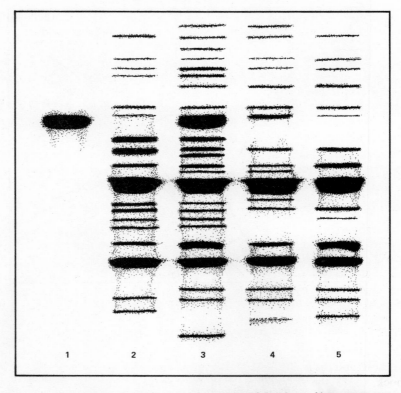

Figure 6.5 Synthesis of polygalacturonase during normal ripening and in mutant tomatoes. Wall proteins were extracted, fractionated by SDS-polyacrylamide gel electrophoresis and stained with coomassie blue to show the amount of protein present in each band. Lane 1, purified polygalacturonase; lane 2, green extract; lane 3, normal ripe fruit; lane 4, neverripe mutant; lane 5, 'ripening inhibitor' mutant.

induces ripening changes, including PG synthesis. Alterations in mRNA metabolism during fruit development and ripening have been studied by analysis of mRNA *in-vitro* translation products by gel-electrophoresis. Several mRNAs which are present in immature green fruit disappear at the mature-green stage but many of those remaining persist throughout the ripening period. There are also about six major translatable mRNAs which appear or increase greatly in quantity during ripening. Analysis of cDNA clones of ripening-specific mRNAs suggests that many more less abundant mRNAs are also involved.

Polygalacturonase mRNA has been identified in RNA extracts from tomatoes by immunoprecipitation of the *in vitro* translation product with antibody raised against the purified enzyme. PG mRNA is not present in green fruit but accumulates during normal ripening *after* the start of ethylene synthesis. Other mRNAs which increase may code for enzymes of ethylene synthesis. When ethylene is supplied to unripe fruit it causes accumulation of ripening-specific mRNAs, including that for PG. However, *rin* tomatoes fail to make PG mRNA even when given ethylene. Kinetic studies show that the ripening-specific mRNAs take 24 hours or more to accumulate in sufficient quantity to be detected by *in vitro* translation. The molecular mechanism for the response is not known.

6.4 Plant responses to stress

Plants show specific responses to many kinds of stress including wounding, infection, ultraviolet light, high temperature, chilling, and anaerobic conditions caused, for example, by waterlogging. These responses are sometimes associated with enhanced ethylene sythesis, which may play a role in activating the production of particular proteins. The molecular basis of these responses has not been completely worked out but there are clear examples of the expressions of specific genes. These include (1) the appearance of a number of proteins (one of which is alcohol dehydrogenase) under anaerobic conditions; (2) the cessation of normal protein synthesis and the production of a group of new proteins in response to heat shock, and (3) the production of a range of enzymes in response to infection or UV light.

Induction of enzymes involved in synthesis of flavenoids and furanocoumarins in response to UV light and infection

There is a relationship between the responses of some plants to UV damage, pathogen attack and perhaps also to wounding. The epidermal cells of some plants accumulate UV-protective materials in the vacuoles when irradiated. Compounds with phytoalexin activity are biochemically related to these and accumulate in response to fungal infection. 'Elicitors' (compounds derived from the pathogen or the host; frequently cell wall fragments) can induce the accumulation of phytoalexins although there is some debate about the physiological relevance of this.

Cell cultures of *Petroselinum hortense* (parsley) produce enzymes of phenylpropanoid metabolism (group I enzymes) and the flavenoid glyco-

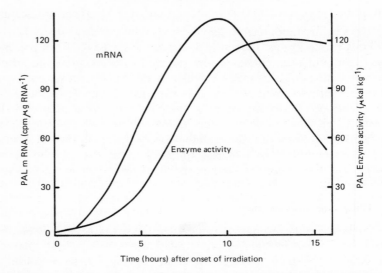

Figure 6.6 Effect of light on phenylalanine ammonia lyase and its mRNA. Parsley cell cultures were continuously irradiated with UV light. Phenylalanine ammonia lyase was measured by enzyme assay and the mRNA by *in-vitro* translation and immunoprecipitation of the translation product with specific antibody. (Redrawn from Schröder *et al.*, 1979.)

side pathway (group II enzymes) in response to UV light (Hahlbrock *et al.*, 1982). There are three enzymes in group I, including phenylalanine ammonia lyase (PAL) and 13 in group II, including chalcone synthase (CHS). PAL and the other enzymes in group I are rapidly but transiently induced in response to UV light (Figure 6.6). The levels of the mRNA have been measured by extraction, *in-vitro* translation and immunoprecipitation and also by hybridization to specific cDNA probes. The mRNA for PAL increases rapidly in amount during the induction period. Similar studies with CHS show that the synthesis of its mRNA follows slightly later. The results of experiments using a cDNA probe to measure the rates of transcription of PAL genes in isolated parsley cell nuclei indicate that the accumulation of the mRNA after induction is due to increased transcription. PAL mRNA also accumulates in response to an elicitor preparation from fungal cell walls but CHS does not respond.

   There appear to be several PAL isoenzyme forms, and studies with cultured *Phaseolus vulgaris* (bean) cells have shown that the constitutive level of PAL is due predominantly to enzymes with a high K_m. Upon induction with elicitor, more active forms of the enzyme with a low K_m are synthesized preferentially. The regulation of PAL therefore seems to involve the effects of several 'inducers' on a multigene family.

6.5 Conclusions

There is an almost endless list of stages in plant development which involve the control of gene expression. Sometimes these processes are triggered by specific environmental or hormonal factors as discussed above but in other examples such as flowering in day-neutral plants, the transition from the juvenile to the adult phase in others, or the start of storage protein synthesis during grain-filling, the regulatory factors are less obvious. Even when the identity of specific 'controlling' factors has been established it is clear that we do not know yet what links perception and response. To complicate matters further, the controlling factors only have their effects on particular tissues at specific stages in development. That is to say, during development plant cells differentiate biochemically and become *competent* to respond to particular signals. The molecular mechanisms underlying this aspect of development merit much greater attention.

CHAPTER SEVEN

PLANT AND BACTERIAL GENE EXPRESSION IN NITROGEN-FIXING LEGUME ROOT NODULES

One factor known to limit the growth of higher plants is the availability of nitrogen in the soil. Gaseous nitrogen comprises four-fifths of the earth's atmosphere but the ability to utilize directly this essential component of many biological molecules is restricted to a few groups of prokaryotes. Some leguminous plant species have overcome this problem by developing a highly organized association with nitrogen-fixing bacteria of the genus *Rhizobium*. These soil-borne organisms invade the roots of co-operative plants and become intracellular 'organelles', called bacteroids, which convert atmospheric nitrogen into ammonia for assimilation by the plant host. The plant produces a specialized organ, the root nodule, to house bacteroid cells and provides the appropriate environment and nutrients to support nitrogen fixation. In agronomic terms, the potential benefit of nitrogen-fixing organisms is enormous since it is estimated that they fix about 2×10^8 tonnes of nitrogen per year.

Although the association between *Rhizobium* and its plant host is often described as symbiotic, some biologists believe that *Rhizobium* was originally pathogenic to plants which subsequently evolved to capitalize upon the nitrogen-fixing capacity of rhizobial cells (Vance, 1983). Whatever its origin, the association is of great interest because it involves co-ordinated gene expression and cellular differentiation in both partners.

This chapter will briefly describe the infection process and establishment of functional root nodules (Meijer and Broughton, 1983) and summarize current knowledge of the specific bacterial and plant genes involved in nodule development.

7.1 *Rhizobium* infection of legume roots

The prelude to infection of legume roots by *Rhizobium* is bacterial growth in the rhizosphere. The first stage is contact between rhizobial cells and susceptible root hairs of the host plant. Successful infection only proceeds following contact between compatible plant/*Rhizobium* combinations. In general, the host-range of particular *Rhizobium* species is limited. Thus, *Rhizobium trifolii* infects clover, *R. phaseoli*, bean and *R. leguminosarum*, pea. There are some exceptions to this; certain strains of *R. trifolii* can nodulate peas and *R. leguminosarum* strains, clover. *Rhizobium* species and strains are classified according to host-range and placed in specific cross-inoculation groups. Some of the determinants of host-range are known to be encoded by *Rhizobium* genes.

Rhizobial cells adhere to plant root hairs in a polar fashion; that is, one end of the bacilliform bacterial cell contacts the root hair. The specificity of binding is thought to be mediated by interaction between plant lectins and the bacterial cell wall. Lectins are glycoproteins, synthesized by plant cells, which exhibit high binding affinity and specificity for sugar residues. A lipopolysaccharide component of the *Rhizobium* cell wall has been identified as a possible receptor for lectin binding and there is evidence suggesting that some *Rhizobium* species stimulate the secretion of plant lectins by producing certain oligosaccharides. The genetically-determined host-range of *Rhizobium* species may also be influenced by the nature of particular plant lectins encoded by the plant genome.

In response to the attachment of rhizobial cells, root hairs curl, possibly as a result of a stimulation of phytohormone production. The root hair cell wall invaginates at the point of contact and extends within the cell to produce an infection thread. It has been suggested that the process of cell wall loosening and extension is effected by pectolytic enzymes or phytohormones secreted by bacterial and/or plant cells. The root-hair cell nucleus migrates to the infection thread tip and becomes enlarged; this may be indicative of increased transcriptional activity. As the infection thread progresses through epidermal cells into the outer cortex of the root, cortical cells begin to divide at the onset of nodule formation. Rhizobial cells also divide within the infection thread which branches and penetrates many cortical cells. Bacteria are discharged into the cortical cells through the infection thread tip. On discharge, they are enveloped in the plant cell plasmamembrane and deposited, but compartmentalized, within the cytoplasm of cortical cells. Bacterial cells divide, to a lesser or greater extent depending upon the species, within the envelope and this is accompanied by an increased level of membrane synthesis by the plant. Enlarged infected cells eventually become packed with bacteria (Figure 7.1). Cortical cell division ceases when bacteria differentiate into nitrogen-fixing bacteroids, a developmental step by the bacteria which is not necessarily irreversible.

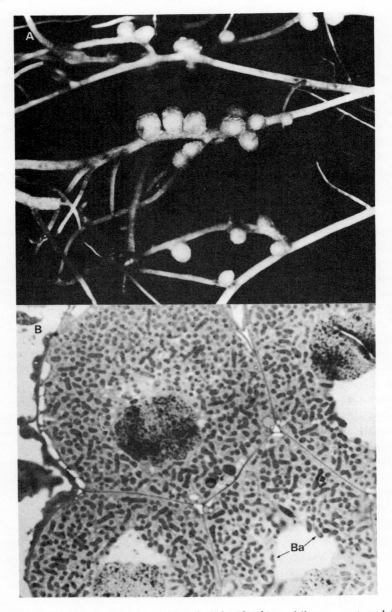

Figure 7.1 Legume root nodules infected with *Rhizobium*: (*A*), pea root nodules; (*B*), nodule cells packed with bacteroids (Ba).

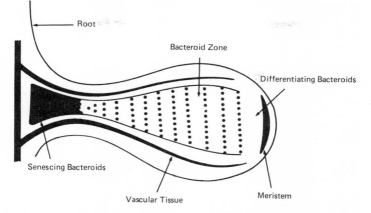

Figure 7.2 Diagrammatic representation of a section through an apical root nodule showing the various zones.

The differentiation of both the bacterial and plant cells is accompanied by increased levels of synthesis of specific RNAs and proteins and other compounds required in the establishment of functional root nodules.

Although the development of root nodules on legume plants is influenced by several physiological factors (temperature, light, soil water content, etc.), the genetic determinants of nodule shape, size and number are expressed by the plant genome in a developmentally-regulated manner in response to penetration by the rhizobial infection thread. Nodules exhibit a variety of morphologies depending on the plant species upon which they develop. Typically, *Pisum, Trifolium* and *Phaseolus* species produce cylindrical apical nodules (Figures 7.1 and 7.2). A root outgrowth develops from meristematic cell division originating in the root cortex. Behind the nodule meristem is a region of non-dividing, differentiating cells containing developing bacteroids and the bulk of the nodule, comprising the bacteroid zone, contains fully-developed nitrogen-fixing bacteroids. The organ is penetrated with vascular tissue which supplies the substrates for nitrogen-fixation, derived from plant photosynthate, and carries away its products.

7.2 Nitrogen-fixing nodules

The biochemistry of nitrogen fixation has been studied in detail in several free-living nitrogen-fixing organisms and *Rhizobium. Rhizobium* species are obligate-aerobic heterotrophic nitrogen-fixers and can also be cultured free of their plant host. However, some of the special biochemical requirements for active nitrogen fixation by *Rhizobium* are met by the host plant in the bacteroid/nodule association.

The reduction of molecular nitrogen (dinitrogen) to ammonium ions (NH_4^+) is a highly energy-demanding conversion. The non-biological, industrial reduction (or fixation) of nitrogen by the Haber process requires very high temperature and pressure together with a catalyst. In nitrogen-fixing organisms, conversion of N_2 to NH_4^+ consumes a great deal of energy (in the form of ATP) and requires a powerful reductant donating electrons of high redox potential. The reaction expressed stoichiometrically is:

$$N_2 + 6e^- + 12ATP + 12H_2O \rightarrow 2NH_4^+ + 12ADP + 12Pi + 4H^+.$$

The enzyme that catalyses this reaction is nitrogenase. Nitrogenase is competitively inhibited by NH_4^+ and inactivated by oxygen. Oxygen is required by bacteroids for oxidation of plant photosynthate to generate the high levels of ATP consumed by the nitrogen-fixation process. The availability of oxygen to bacteroids is regulated by the plant which must ensure an adequate supply for respiration whilst preventing oxygen-inactivation of nitrogenase, by a process known as respiratory protection.

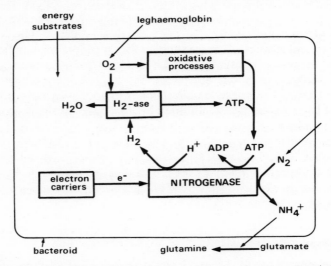

Figure 7.3 Simplified scheme of the biochemistry of nitrogen fixation in a Hup$^+$ *Rhizobium* strain.

Finally, NH_4^+ is removed from the reaction centres preventing feedback-inhibition of nitrogenase activity.

A generalized scheme illustrating bacteroid nitrogen-fixation is shown in Figure 7.3. Electrons pass from a reduced electron carrier (ferredoxin or flavodoxin) to the first component of the nitrogenase complex, called dinitrogen reductase, which then reduces the second component, dinitrogenase, which in turn converts N_2 to NH_4^+. Nitrogenase has a secondary activity, concomitant with nitrogen reduction, in that it also reduces protons to molecular hydrogen. This apparently useless activity consumes energy and generates hydrogen which also competitively inhibits nitrogenase. Some species and strains of *Rhizobium* (designated *Hup*$^+$) can utilize the hydrogen liberated in this reaction by combining it with oxygen catalysed by another enzyme, called hydrogenase. This activity regenerates some of the ATP lost in proton reduction and probably helps remove oxygen from the vicinity of nitrogenase.

The oxygen concentration in bacteroid-containing nodule cells is regulated by a plant protein called leghaemoglobin (Lb). Lb is a myoglobin-like protein with a haem prosthetic group. Like the haemoglobin of animal cells, Lb has a high affinity for oxygen. High levels of Lb are found in the cytoplasm of root nodule cells where it can constitute 20–30% of the cell protein. It was originally supposed that Lb chelated cellular oxygen and so protected nitrogenase in the bacteroids from oxygen inactivation. The current view is that Lb acts as an oxygen-carrier which buffers the cell environment from wide fluctuations in oxygen concentration and ensures adequate supplies for both host cell and bacteroid oxidative processes. The synthesis of Lb is specific to nitrogen-fixing nodule cells of leguminous plants and the plant Lb genes are activated and expressed only in response to *Rhizobium* infection. The haem prosthetic group of Lb is, in contrast, synthesized by rhizobial cells so the production of functional Lb is regulated by both partners.

Ammonium ions, resulting from the reduction of nitrogen by nitrogenase, pass from the bacteroid cells to the nodule cell cytoplasm where they are incorporated into glutamine by glutamine synthetase and thence to other amino acids and nitrogen-containing compounds for redistribution within the plant.

7.3 Location of genetic determinants in nodule formation

The development of functional nodules is a progression of stages involving the coordinated expression of a number of bacterial and plant genes. The isolation of mutants defective in specific functions required in the establishment of nitrogen-fixing nodules has been useful in identifying the genetic contribution of each partner. For example, there are mutations which prevent initial infection and others which result in abortive infection threads and so the nodule-forming ability is lost. Mutations of both plant and bacterial genomes have been identified which permit nodule formation but lead to ineffective nodules in that they do not fix nitrogen. The steps thought to be controlled by *Rhizobium* genes (Vincent, 1980) include recognition of host plants, root hair curling, the formation of infection threads, release of bacteria from the infection thread, nodule development and bacteroid differentiation in addition to the nitrogenase structural and related genes involved in nitrogen fixation.

7.4 *Rhizobium* genes

The genetics of several *Rhizobium* species has been studied in detail. Mutant strains of *Rhizobium* have been isolated, either from natural populations or by *in-vitro* mutagenesis, that are defective in certain symbiotic functions. Some mutant strains infect plants and induce nodule formation but do not fix nitrogen. These are designated nod^+fix^- strains; those that do not elicit nodule formation at all are denoted nod^-. Since the establishment of functional nodules is believed to involve at least ten distinct steps (Vincent, 1980), controlled genetically by *Rhizobium*, abolition of a function specifying any one of these steps following mutation of one or more *Rhizobium* genes is likely to lead to a breakdown in nodule development. With a complete series of different mutations, it should be possible to identify each step and then, eventually, to map particular functions on the DNA molecule encoding them.

It is now known that many of the bacterial genes involved in the establishment of nitrogen-fixing nodules, at least in the fast-growing *Rhizobium* species (*R. trifolii, R. leguminosarum, R. phaseoli, R. meliloti*), are located on one of several extrachromosomal giant plasmids found in these organisms (Nuti *et al.*, 1977). Some mutant rhizobia which do not nodulate plants or fix nitrogen (nod^-fix^- mutants) were found to be deficient in part of their normal plasmid complement. Incubation of wild-type ($nod^+ fix^+$) *R. leguminosarum* at elevated temperature resulted in plasmid loss (known as curing) and subsequent inability to nodulate peas. *Rhizobium* plasmids can be transferred between species or strains by transconjugation, transduction or transformation (Johnston and Beringer, 1982) and transfer of plasmid DNA from nod^+fix^+ organisms to mutant strains lacking these functions resulted in the mutants regaining the ability to nodulate plants and fix nitrogen. In similar transfer experiments it has been shown that the plant host-range of different *Rhizobium* species is also a plasmid-borne trait (Johnston *et al.*, 1978; Downie *et al.*, 1983a). Thus *R. phaseoli*, which normally nodulates only bean, could subsequently nodulate pea plants when a plasmid of *R. leguminosarum* (a pea nodulator) was transferred to a plasmid-cured strain of *R. phaseoli*. Genes encoding the hydrogenase system, in those organisms that possess it (Hup^+ strains), are located on plasmid DNA too.

Purification of *Rhizobium* plasmids is a relatively straightforward procedure (Nuti *et al.*, 1982) and their structure and function can be studied by recombinant DNA techniques, restriction endonuclease mapping and other methods currently used in molecular biology. An example which illustrates this well is the characterization of a plasmid (pRL1JI) of *R. leguminosarum* (Hirsch *et al.*, 1980). pRL1JI is an extrachromosomal circular DNA molecule 200 kilobase-pairs in size. From plasmid transfer experiments it was found that pRL1JI encoded the genetic determinants for nodulation, nitrogen fixation and host-range specificity in pea. Several subfragments of pRL1JI were cloned into a cosmid vector which replicates both in *E. coli* and *Rhizobium*. A series of overlapping sub-fragments of pRL1JI were separately introduced into a strain of *R. leguminosarum* which had been cured of its nodulation plasmid and two DNA fragments with about 10 kilobase-pairs in common were found to specify nodulation and host-range following transfer, suggesting that these two functions are very closely linked on the plasmid DNA. Further mapping was carried out by inserting a bacterial transposon, Tn5, into multiple sites within pRL1JI, thus inactivating specific *Rhizobium* genes at the sites of insertion. Analysis of Tn5-induced mutations by restriction enzyme and Southern hybridization mapping revealed the locations of inserted DNA. Various mutant clones of the *Rhizobium* plasmid, containing Tn-5 insertions, were tested for their ability to nodulate peas and fix nitrogen. Tn-5 insertions at some sites resulted in loss of the nitrogen fixation function whilst at other sites, the nodulation ability was lost. Thus, the position of a mutation in the plasmid DNA was correlated with loss of a particular function (Figure 7.4).

The gene encoding one protein of the nitrogen fixation enzyme complex was identified on restriction fragments of pRL1JI by Southern hybridization using a DNA probe containing the *nif* KDH coding sequence of the free-living nitrogen-fixing bacterium *Klebsiella pneumoniae*, which is homologous with that of *Rhizobium* (Downie *et al.*, 1983b).

The findings from these experiments show that the *nod, fix* and host-

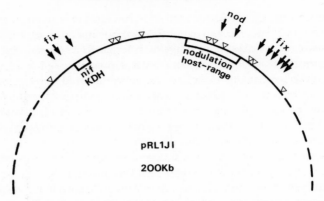

Figure 7.4 Physical and genetic map of the *R. leguminosarum* symbiotic plasmid pRL1JI involved in nitrogen fixation and nodulation. Arrows denote the sites of transposon (Tn5) insertions which block *nod* and *fix* functions. A region of homology with a nitrogenase structural gene (*nif* KDH) of *Klebsiella pneumoniae* is close to the nodulation and host-range genes. *Bam* H1 restriction enzyme sites ($\bigtriangledown$).

range genes on the *R. leguminosarum* plasmid pRL1JI are located in a sequence encompassing only 45 kilobase-pairs of DNA (Figure 7.4). So perhaps fewer *Rhizobium* genes are required for nodulation than were originally supposed. One future goal will be to determine the nucleotide sequences encoding *nod* and *fix* functions and to characterize their RNA transcripts and protein products in order to gain information about the mechanisms controlling their expression.

7.5 Plant genes involved in nodulation

Although some mutations in the plant genome have been identified which affect nodule formation, most studies of the plant's contribution to nodulation have been directed to an analysis of novel plant-encoded gene products found in nodule cells. Leghaemoglobin (Lb), which regulates the nodule oxygen levels, is the major plant gene-coded protein in bacteroid-containing nodule cells. Lb has a haem prosthetic group and an apoprotein component of 15 600 to 15 900 molecular weight. Four distinct variants of Lb (Lba, Lbc_1, Lbc_2, Lbc_3) have been identified; each differs slightly in amino acid sequence. Polyadenylated RNA, isolated from nodule cells, has been translated *in vitro* to produce Lb, as shown by immunological methods. In fact, the messenger RNA encoding Lb was one of the first plant mRNAs to be identified (Verma *et al.*, 1974). *In-vitro* generated cDNA copies of Lb mRNA have been used in hybridization studies to determine the location of Lb genes in soybean (Baulcombe and Verma, 1978). These have shown that there are about 40 copies of the Lb genes per haploid genome. Individual Lb genes have been isolated from a cloned DNA library of the soybean genome and structural analysis has demonstrated that each of the four Lb variants is encoded by a different gene. Thus, the population of Lb molecules in nodule cells is derived from a multigene family. The structures of two Lb genes (Lba and Lbc) have been determined by nucleotide sequencing, which showed that the protein coding sequences (exons) of each gene are interrupted by three non-coding sequences (introns) (Hyldig-Nielsen et al., 1982) (Figure 7.5). Although the coding sequences in each gene are of similar size, the intervening sequences are not and so the overall lengths of the two genes are different.

An analysis of the amino acid sequence of Lb has revealed several regions of homology with animal globin and, since both proteins have similar functions, it is likely that they are related in evolutionary terms. It is thought that the amino acid sequence encoded by the central exons of the Lb gene is the haem prosthetic group binding domain. Non-functional Lb pseudogenes have also been isolated: two are truncated (Brisson and Verma, 1982) and another has enlarged intervening sequences (Wiborg et al., 1983).

Expression of Lb genes appears to be controlled by nucleotide sequences

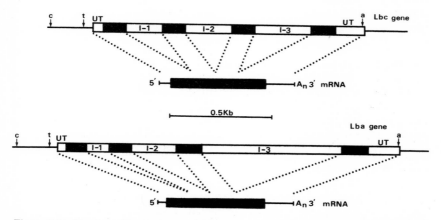

Figure 7.5 Structure of two leghaemoglobin genes and their mRNAs from soybean. Introns (I1-3), protein-coding exons (▬) and the mRNA untranslated regions (UT) are shown together with putative transcription control sequences (c, CAT box; t, TATA box; a, poly(A) signal sequence).

characteristic of other plant and eukaryotic genes (Chapter 3). Upstream of the 5'-end of the Lb genes there are eukaryotic promoter-like sequences CCAAG and TATAA (the 'CAT' and 'TATA' boxes) thought to define the initiation of transcription. A sequence GAUAAA, which probably specifies poly(A) addition and related to the consensus eukaryotic poly(A) signal sequence AAUAAA, is located 20 nucleotides from the 3'-end of Lb mRNAs.

In addition to Lb, there are at least 20 other polypeptides encoded by the plant genome and found only in root nodules elicited by *Rhizobium*. These proteins are known collectively (including Lb) as 'nodulins' and have been studied in detail in soybean roots infected by *Rhizobium japonicum*. Fuller *et al.* (1983) have subdivided nodulins into three groups based upon their likely functions. Group 1 nodulins are nodule structural components; group 2 nodulins are enzymes responsible for assimilation into the plant of nitrogen fixed by bacteroids; group 3 nodulins are proteins required for supporting bacteroid functions.

Although specific functions have not yet been assigned to particular group 1 nodulins, a number of polypeptides which could belong to this group have been detected in nodule membrane fractions and might be involved in transport of metabolites across the membrane enclosing the bacteroids.

Two group 2 nodulins have known functions: one is a nodule-specific form of glutamine synthetase (Cullimore *et al.*, 1983) required for the assimilation of ammonium ions, a primary product of bacteroid nitrogen fixation, into amino acids for consumption by the plant. A second group 2 nodulin is a polypeptide called nodulin-35 (mol. wt. 35 000) and is second in abundance only to Lb in soybean nodules. The enzyme uricase consists of nodulin-35 polypeptide subunits (Bergmann *et al.*, 1983) and catalyses the breakdown of uric acid (a catabolite of purine bases) to allantoin. Uricase is present in the peroxisomes of uninfected nodule cells which are adjacent to cells containing *Rhizobium* bacteroids. It is thought that allantoin and allantoic acid produced in these uninfected cells are the major stored or transported forms of nitrogen fixed in bacteroid-containing soybean nodule cells. Lb belongs to the group 3 nodulins.

In-vitro translation following hybrid-selection with cDNA clones of soybean nodule polysomal polyadenylated RNA has been used to study four further nodulins of unknown function. Nodulin-44, a 44 000 molecular weight polypeptide also called nod A, is the translation product of a 1.6kb mRNA. Similarly, nodulin-27 (nod B), nodulin-24 (nod C) and nodulin-100 (nod D) mRNAs are 1.15kb, 0.77kb and 3.15kb in size respectively. Each bacteroid-containing nodule cell contains about 600 000 molecules of polysome-associated mRNA and hybridization studies have shown that, of these, some 90 000 (15%) are Lb mRNA molecules, about 36 000 (6%) are nodulin-44 mRNAs and 3000–7000 (0.5%–1.1%) are mRNAs each of nodulins-27, -24 and -100. Because the nodulins and their mRNAs are present in variable copy numbers, this suggests that the half-life of each nodulin mRNA is not the same and is related to differences in the rates of transcription of nodulin genes or to rates of mRNA degradation. The factors controlling this are not yet understood.

A further interesting feature concerns the role of *Rhizobium* in activating nodulin genes at the onset of nodule formation. Nodulin mRNAs, including that for Lb, have been detected in soybean roots 5 days after inoculation with *Rhizobium* (Fuller and Verma, 1984) and before nodules are visible on the roots. Nodulins and their mRNAs accumulate during nodule development and reach a steady-state level 10–12 days after root inoculation (7–9 days after infection is established). However, nitrogenase activity is first detectable some 3–5 days after the appearance of the plant gene products, suggesting that bacteroid gene expression is not required for the activation of nodule-specific plant genes (Figure 7.6). Nevertheless, the early infection events must influence the induction of plant nodulin genes.

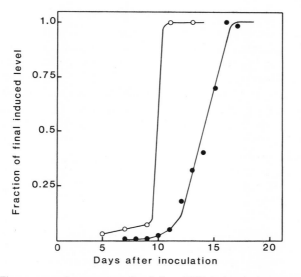

Figure 7.6 Time-course of appearance of nodulin mRNA (○) and nitrogenase activity (●) in soybean root nodules. (Redrawn from Fuller and Verma, 1984.)

In nodules ineffective in nitrogen fixation, elicited by mutant strains of *Rhizobium* (*nod$^+$ fix$^-$* mutants), Lb mRNA and protein is still synthesized albeit at a somewhat reduced level which has been interpreted to mean that the bacteria can influence the level of expression of plant nodulin genes (Verma *et al.*, 1981). Presumably, early steps in the invasion process are responsible for stimulating the activity of plant genes and this might occur first in the root-hair cell nucleus which becomes enlarged soon after penetration by the infection thread. Whether rhizobial 'activator' molecules switch-on individual plant genes or initiate the programmed expression of a bank of nodule-specific plant genes is not yet known. Each of the nodulin genes is probably regulated independently because, for Lb at least, the different polypeptide varients produced by expression of this multi-gene family accumulate at different rates during nodule development.

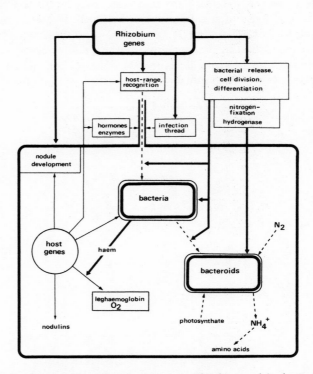

Figure 7.7 Summary of the interactions between *Rhizobium* and its host plant in the development of nitrogen-fixing root nodules.

7.6 Prospects

Important details of the molecular processes which lead to the development of nitrogen-fixing root nodules in leguminous plants are beginning to emerge (Figure 7.7). Future research will, no doubt, be aimed towards further characterization of both *Rhizobium* and plant genes, the mechanisms regulating their expression and the function of proteins synthesized during infection and nodule development. At present, little is known about the interaction between the bacterial and host genomes at the molecular level or how specific genes are expressed in a coordinated manner in both organisms.

Only when these processes are understood in detail will it be possible to consider, realistically, transferring the nitrogen-fixing capability to non-leguminous plants (see Chapter 10).

CHAPTER 8

GENETIC COLONIZATION OF PLANTS
BY *AGROBACTERIUM*

Agrobacterium tumefaciens causes crown gall disease in plants. Crown galls
are cancerous growths composed of disorganized and proliferating plant
cells. *Agrobacterium* is an oncogenic agent which neoplastically transforms
normal plant cells into tumour cells and in so doing, directs the plant to
synthesize special nutrients that support growth of the bacteria around the
tumour. A distinct competitive advantage is conferred upon the inciting
Agrobacterium strain because only that, or a related strain, can utilize the
nutrients produced by the tumour as an energy source. The neoplastic
transformation process involves transfer of genetic information from the
bacteria to the plant and so can be considered as a type of genetic
colonization (Schell *et al.*, 1979).

Crown gall and related diseases have been extensively studied over the
past 70 years or so and this is currently one of the most active areas of
research in plant molecular biology. The reasons for this are firstly that it
might lead to a better understanding of oncogenesis and secondly that
Agrobacterium is probably the best system we have at present for
genetically manipulating plants by molecular means (see Chapter 10).

8.1 *Agrobacterium* and plants

Members of the genus *Agrobacterium* are soil-borne, Gram-negative
bacteria. Four distinct species are recognized: *A. tumefaciens* causes gall
formation (disorganized undifferentiated cells) (Figure 8.1) on plants of at
least 90 different families of dicotyledonous angiosperms and gymnos-
perms. Some strains of *Agrobacterium* do not produce galls but induce
growth of teratomas, which are tumours containing abnormal differentiated cells, often with the appearance of shoots or roots. *A. rubi* produces
small gall-type tumours on the stems of raspberry and a few other plants. *A.
rhizogenes* is the causative agent of hairy root disease often found on

Figure 8.1   Crown gall tumour on the stem of a tomato plant.

nursery stock and *A. radiobacter* is an avirulent species in that it does not normally incite tumour formation on plants.

Although *Agrobacterium* species (and *A. tumefaciens* in particular) infect most dicotyledonous plant species, the host-range of some bacterial strains is limited. Both plant and bacterial factors are thought to be involved in determining host-range but these are not well understood. Monocotyledonous plants are resistant to infection by *Agrobacterium*, possibly because they lack cell wall components necessary for recognition and binding of bacteria, although it is likely that other features also contribute to this resistance.

8.2 Infection and tumour growth

Agrobacterium will infect a susceptible plant and induce tumour development only at the site of a wound. It appears that *Agrobacterium* binds to pectic material in plant cell walls rather than to lectins which have been implicated in the binding of *Rhizobium* to legume root hair cells (see Chapter 7). Shortly after the early infection events, *Agrobacterium* causes localized cell proliferation leading to the development of a plant tumour. Many of the intervening steps between infection and tumorigenesis have not yet been characterized. *Agrobacterium*-induced tumours can be removed from the plant and cultured *in vitro* where they continue to grow indefinitely. A series of now classic experiments by Braun and co-workers in the 1940s (see Braun, 1982) clearly demonstrated that once tumour formation had been initiated, the further presence of *Agrobacterium* cells was not required for subsequent tumour proliferation. Thus, plant cells neoplastically transformed by *Agrobacterium* will proliferate indefinitely as a stable tumour cell-line in the absence of the inciting organism. This finding led Braun to argue that *Agrobacterium* secretes a tumour-inducing principle (TIP) the activity of which causes plant cell transformation. It was many years later that the nature of the TIP was finally elucidated.

A second important feature of tumour cells is that their growth in culture does not require the presence of plant hormones in the culture medium. This contrasts with normal plant cells in culture which usually require auxin and cytokinin to maintain growth and viability. The hormone-independent growth autonomy of *Agrobacterium*-induced tumour cell lines suggests that significant changes in the internal hormone levels are a direct consequence of the transformed state. Changes in hormone balance are also implicated in the generation of teratomas elicited by some *Agrobacterium* strains. Teratomas on plants and in culture can consist predominantly of green shoot tissue or in other cases root tissue. Tumours cultured from gall tissue are typically amorphous masses of relatively undifferentiated cells.

Under the appropriate conditions, cultured cells derived from healthy plants of certain species will differentiate and eventually develop into a complete plant. Regeneration of plant tumour cells is more difficult to achieve. Tobacco teratomas have been grafted onto healthy tobacco plants and eventually the tumourous parts of the plant produced fertile flowers. Similarly, tobacco and sunflower teratomas, resulting from infection by attenuated *Agrobacterium* strains, have been regenerated into fertile plants which still had many characteristics of the transformed state. Seed produced by these plants reverted to the non-transformed state. This means that the factors responsible for maintaining the tumourous condition of such plants are not transferred, in general, through meiosis. (For exceptions to this, see Chapter 10.)

8.3 Opines

Plant tumours resulting from *Agrobacterium* infection synthesize a variety of unusual amino acid derivatives called opines (Petit *et al.*, 1970). Opine synthesis is a unique characteristic of tumour cells; normal plant tissues do not produce these compounds in any significant quantity. Originally, two groups of opines were recognized: the octopine family are carboxyethyl-derivatives of arginine; nopaline and related compounds are dicarboxypropyl-derivatives of arginine (Figure 8.2). Two further groups of opine have been identified more recently. Agropine is a bicyclic sugar derivative of glutamic acid and agrocinopines are phosphorylated sugars.

Opines synthesized in tumour cells can be catabolized by *Agrobacterium* to generate energy. Octopine, for instance, is converted to arginine and pyruvic acid by bacterial enzymes and assimilated into the bacterial cell metabolism. Particular *Agrobacterium* strains induce tumours which then produce a specific opine. That opine can then be catabolized only by the inciting bacterium or a related strain. For example, *A. tumefaciens* strain C58 incites the formation of nonapline-producing tumours; strain C58 can also catabolize nopaline. Strain Ach 5 causes tumours that synthesize octopine and can utilize octopine but not nopaline. The generation of tumours producing specific opines catabolizable only by the inciting *Agrobacterium* strain is a central feature of the pathogenic relationship between bacterium and plant. This proved to be a useful phenotypic characteristic which contributed to an understanding of the genetic basis of *Agrobacterium*-induced tumorigenesis.

8.4 Tumour-inducing (Ti) plasmids

As we have seen, the concept of the tumour-inducing principle (TIP) of *Agrobacterium* developed from the observation that tumour cells will proliferate in the absence of the inciting organism. In an attempt to explain the phenomenon, several theories were originally proposed, often not supported by hard evidence. A postulated chemical messenger (e.g. protein or RNA) secreted by *Agrobacterium* which then activated certain latent plant genes was not isolated or identified. Other claims that viruses or bacterial DNA induced tumorigenesis were not substantiated experimentally either. However, the techniques available at this time to detect transferred nucleic acid were relatively insensitive compared with some of the more recently developed hybridization methods (Chapter 1).

The problem was finally resolved after observations made in the late

1960s and early 1970s. It was found that by incubating a virulent strain of *A. tumefaciens* (C58) at high temperature (36°C), virulence was permanently lost (Hamilton and Fall, 1971). Laboratories in the United States and Europe provided strong evidence implicating the involvement of bacterial extrachromosomal genetic elements in tumorigenesis. Following mixed inoculation of plants with virulent and avirulent *A. tumifaciens* strains, it was observed that the virulence trait was transferred to the avirulent strain (Kerr, 1969). Analysis of bacterial DNA showed that many *Agrobacterium* strains possess one or more giant extrachromosomal plasmids with molecular weights of $100-150 \times 10^6$ (Zaenen *et al.*, 1974). The loss of virulence of strain C58 following heat treatment was correlated with loss of a plasmid, thus establishing a direct link. Also, plasmid transfer from a virulent bacterium to an avirulent plasmid-lacking strain conferred oncogenicity (the ability to cause tumour formation) upon the recipient strain. The conclusion of these experiments (Van Larabeke *et al.*, 1975; Watson *et al.*, 1975), and the currently accepted view, is that plant cell transformation is caused by an *Agrobacterium* plasmid DNA molecule now known as the Ti (tumour-inducing) plasmid. In addition to the Ti-plasmid, some *Agrobacterium* strains contain other large plasmids apparently not involved in tumorigenesis.

8.5 Neoplastic transformation of plant cells by the Ti plasmid

The identification of the Ti plasmid as the TIP marked the beginning of a new phase in *Agrobacterium* research. One obvious question asked was how the Ti plasmid induces tumour formation. It was known that conversion of normal animal cells to a tumorous state by certain viruses was associated with integration of viral DNA into host cell chromosomes. Does the Ti plasmid integrate into plant cell DNA? Initial attempts to detect Ti plasmid DNA in plant tumour cell chromosomes by DNA renaturation experiments gave disappointingly negative results. However, when cloned restriction enzyme-generated sub-fragments of Ti plasmid DNA were used individually to probe tumour cell DNA, two contiguous Ti DNA fragments gave positive hybridization signals with nuclear but not with organelle DNA (Chilton *et al.*, 1977). From these and other results, it became clear that a small segment (approximately one-tenth) of the Ti plasmid DNA was integrated into the nuclear DNA of plant tumour cells. The fragment of Ti plasmid DNA which becomes integrated is known as T-DNA.

8.6 Structure of Ti plasmids

The capacity of different *A. tumefaciens* strains to induce tumours to synthesize certain opines has been correlated with the presence of specific Ti plasmids. Ti plasmids also specify catabolism of a particular opine. Thus, the nopaline plasmid of *A. tumifaciens* strain C58 encodes the genetic determinants for nopaline synthesis and catabolism together with those conferring oncogenicity.

Both nopaline and octopine Ti plasmids have been studied in detail and the cutting sites of several restriction enzymes mapped on each circular DNA molecule (Figure 8.3). The nopaline plasmid pTiC58 is about 194 000 base pairs in size (Depicker *et al.*, 1980) whilst the octopine plasmid pTiAch 5 is a little longer: 213 000 base pairs (DeVos *et al.*, 1981). Hybridization experiments with cloned fragments of each DNA have revealed several regions of homology between octopine and nopaline plasmids. These probably delimit DNA sequences with functions common to both plasmids. However, the octopine and nopaline plasmids are distinct and contain long segments of non-homologous DNA (Figure 8.3).

The T-DNA region of the nopaline Ti plasmid pTiC58 which becomes integrated into the host chromosomes is about 23 kilobase pairs long. The corresponding T-DNA of the octopine plasmid pTiAch 5 is about 21

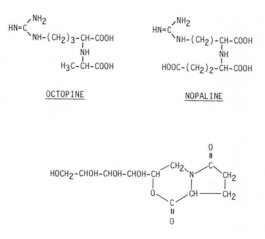

Figure 8.2 Types of opine synthesized in plant cells transformed by *Agrobacterium*.

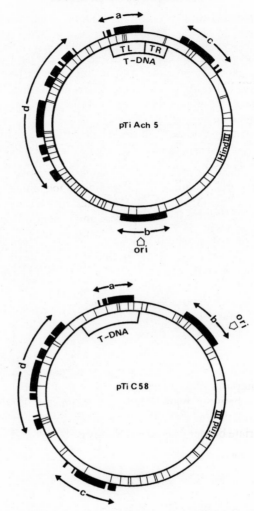

Figure 8.3 Physical maps of octopine (pTiAch 5) and nopaline (pTiC58) Ti plasmids. Regions common to both plasmids are represented as black arcs (*a-d*). The origin of replication (*ori*) and *Hind* III restriction enzyme sites are shown. Octopine T-DNA is composed of two segments, TL and TR.

kilobase pairs long. However, octopine T-DNA, when transferred to the plant cell nucleus, sometimes integrates in two separate pieces. The left-hand piece, TL, is 13 kilobase pairs long and the right-hand piece, TR 8 kilobase pairs. Within the T-DNA of octopine and nopaline plasmids there are 3 regions common to both DNAs (Figure 8.3).

8.7 Organization of integrated T-DNA

Having established that a 20–23 kilobase pair fragment of the Ti plasmid is transferred to the plant cell nucleus, several questions concerning its subsequent organization and distribution within the plant genome have been posed. How many copies are integrated? Does integration occur at specific sites in nuclear DNA? Do T-DNA nucleotide sequences influence specificity of integration? Answers to these and other questions have come from Southern hybridization analysis using cloned Ti plasmid DNA restriction enzyme fragments to probe nuclear DNA isolated from tumour cell lines. Digestion of plant genomic DNA containing integrated T-DNA (of the nopaline plasmid pTiC58), with restriction enzymes, produced a number of fragments with sizes similar to those found by digestion of unintegrated pTiC58 DNA. Additional larger fragments, not found in pTiC58 DNA, were also detected which represented the edges of T-DNA joined to plant DNA. Detailed mapping showed that T-DNA from nopaline plasmids integrated into plant DNA in one continuous piece. In most nopaline tumour cell lines only one or a few copies of T-DNA are integrated per haploid plant genome.

The junctions between plant DNA and integrated T-DNA in several nopaline tumour cell lines have been characterized by nucleotide sequencing the appropriate border fragments (Yadav *et al.*, 1982; Zambryski *et al.*,

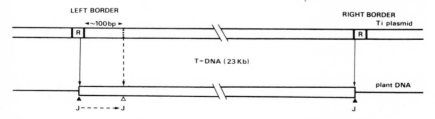

Figure 8.4 Organization of nopaline T-DNA borders in the Ti plasmid (upper line) and after integration into nuclear DNA (lower line). j denotes the junctions between T-DNA and plant DNA. The left T-DNA border is variable by about 100 base pairs in different cell lines (broken arrows) but both borders are close to a 24 base-pair repeat sequence (R) in Ti plasmid DNA.

Nopaline left	G G C A G G A T A T A T T G T G G T G T A A A C
Nopaline right	G A C A G G A T A T A T T G G C G G G T A A A C
Octopine TL-left	G G C A G G A T A T A T T C A A T T G T A A A T
Octopine TL-right	G G C A G G A T A T A T A C C G T T G T A A T T
Octopine TR-left	G G C A G G A T A T A T C G A G G T G T A A A A
Octopine TR-right	G G C A G G A T A T A T G C G G T T G T A A T T

Figure 8.5 The 24 base-pair repeat sequence found close to the T-DNA borders in nopaline (pTiT37) and octopine (pTi15955) Ti plasmids.

1982). The precise sites of integration at the left and right borders are found by seeing where the integrated T-DNA nucleotide sequences diverge from those of the unintegrated T-DNA in the Ti plasmid. A conserved 24-base-pair direct repeat sequence has been found in Ti plasmid DNA close to the T-DNA borders. The right T-DNA border is located 1 or 2 nucleotides from the 24-base-pair sequence. The left T-DNA border is more variable in different nopaline cell lines and is located either within the 24-base-pair sequence or up to 100 nucleotides to the right of it (Figure 8.4).

The organization of integrated T-DNA derived from octopine Ti plasmids is more complex and differs between tumour cell lines. In some cell lines octopine T-DNA, like nopaline T-DNA, integrates as one contiguous 21 kilobase pair segment. In others, the left-hand portion of octopine T-DNA (TL) alone integrates to a level of about 1 copy per haploid plant genome. Yet other cell lines have been isolated with both the left-hand (TL) and right-hand (TR) octopine T-DNA integrated at separate locations in plant DNA. Multiple copies (10 or more per haploid genome) of TR DNA were found to be integrated. A 24-base-pair T-DNA sequence, similar to that at the junctions of nopaline T-DNA, has also been found close to both junctions of the TL and TR pieces of octopine T-DNA (Barker *et al.*, 1983).

The conserved junction sequences (Figure 8.5) must be important in defining integration of particular pieces of Ti plasmid DNA although the precise mechanism which controls this process is not yet known. In contrast to the specificity exhibited in the T-DNA insertion sequences, it appears that no such specificity resides in the plant DNA so far as it is known; T-DNA has been found inserted at several different locations within the plant genome.

8.8 Functions encoded by Ti plasmid DNA

The genetic determinants specifying oncogenicity, opine synthesis and catabolism, and several other functions associated with the neoplastic state of *A. tumifaciens*-induced plant tumours, are located on Ti plasmids. Mapping of Ti DNA sequences containing genes which express these functions has been achieved experimentally using a number of methods. In general, location of sites of mutations on particular pieces of Ti DNA correlated with loss of certain functions has yielded much information about the genetic organization of Ti plasmids. Several naturally-occuring phenotypic variants have been isolated. In addition, a large number of mutants have been generated by insertion mutagenesis using bacterial transposons as well as deletion mutagenesis of Ti plasmid DNA fragments. Insertion and deletion mutagenesis is a useful approach because the precise location of mutations can be mapped with restriction endonucleases.

Mutations at several sites in Ti plasmid DNA result in loss of oncogenicity (*onc⁻* mutants), that is, the ability to induce tumour formation is lost. Some *onc⁻* mutations map within the T-DNA region whilst several others map in a segment to the left of the T-DNA portion of the Ti plasmid. Thus, sequences within the T-DNA confer oncogenicity together with the segment adjacent to it known as the virulence (*vir*) region (Figure 8.6). The virulence region is thought to control the integration of T-DNA into the plant genome.

Ti plasmid genes encoding enzymes responsible for the degradation of opines also reside outside of the T-DNA and in a segment to the right of it. The opine catabolism genes are prokaryotic in nature in that they are inducible operons. Their transcription is specifically activated, by de-repression, in the presence of opine molecules. Ti plasmids specifying a particular opine (octopine, nopaline, etc.) encode those enzymes required for catabolizing the same opine.

Certain functions have been mapped to the T-DNA region which are expressed following its integration. Opine synthesis is controlled by T-DNA and the enzymes octopine synthase and nopaline synthase are encoded on the T-DNA at loci called *ocs* and *nos* on octopine and nopaline plasmids respectively.

The region of T-DNA common to both octopine and nopaline plasmids express certain functions affecting tumour morphology. Mutations in the 'rooty' locus (*roi*) result in tumours developing abnormal roots rather than the usual undifferentiated cells produced by a wild-type Ti plasmid. Adjacent to the *roi* locus is a 'shooty' locus (*shi*); mutations in *shi* result in tumours with abnormal shoot tissue. Clearly, these loci are implicated in controlling hormone levels in tumour cells. If the relative levels of auxin and cytokinin maintain a particular state of cell differentiation, then mutations in the *shi* or *roi* locus which alter this hormone balance are likely

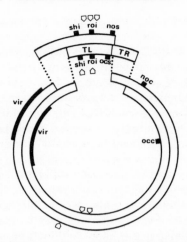

Figure 8.6 Composite functional map of nopaline (outer ring) and octopine (inner ring) Ti plasmids. The non-integrated part of the Ti plasmid contains an oncogenic virulence region (*vir*) and genes coding for catabolism of octopine (*occ*) and nopaline (*noc*). Octopine synthase (*ocs*) and nopaline synthase (*nos*) genes are in the T-DNA and loci controlling 'shooty' (*shi*) and 'rooty' (*roi*) tumour morphology are in regions common to the T-DNA of both plasmids. Open arrows show additional oncogenicity loci.

to lead to proliferation of roots or shoots. The corollary to this is that undifferentiated tumour cells are sustained by the balanced expression of genes at the *shi* and *roi* loci in T-DNA (Nester and Kosuga, 1981; Caplan *et al.*, 1983). It is interesting to note in this respect that oncogenes implicated in tumour induction in anim?: cells are thought to express functions affecting the normal level of growth factors which promote cell proliferation. Functional loci on Ti plasmids are shown in Figure 8.6.

8.9 Transcripts of T-DNA

The expression of T-DNA has been studied by analysis of RNA transcripts isolated from tumour cells (Bevan and Chilton, 1982). It has been estimated that T-DNA transcripts comprise only 0.001% of the tumour cell messenger RNA population. Hybridization experiments have shown that nearly all of the T-DNA is transcribed but some sequences more actively than others.

Northern blot hybridization has been used to determine the size and number of T-DNA transcripts. In nopaline tumour cell lines, 13 polyadeny-lated RNA transcripts have been found ranging in size from about 0.5

kilobases to 3 kilobases. A similar size distribution has been observed in 9 T-DNA transcripts of octopine tumours. Many of the aforementioned transcripts have been mapped to particular regions of the T-DNA using restriction enzyme-generated DNA fragments as hybridization probes and by S1 nuclease mapping (see Chapter 1). This showed that specific RNAs are transcribed from the *nos, ocs, shi* and *roi* loci, and other less well characterized regions of both octopine and nopaline T-DNA (Figure 8.7). Hybridization of RNA with separated strands of T-DNA has also revealed that transcription is bi-directional: some RNAs are transcribed from one DNA strand, others from the second strand.

Transcription of T-DNA in nuclei isolated from tumour cells is sensitive to inhibition by low concentrations of α-amanitin. This strongly suggests that T-DNA is transcribed by host cell RNA polymerase II (see Chapter 3).

It is likely that many if not all T-DNA transcripts are messenger RNAs; most are polyadenylated and have been found associated with polysomes. However, because of their low abundance in tumour cells, it has proved difficult to purify sufficient quantities of most T-DNA transcripts to test their template activity by translation *in vitro*. The RNA transcribed from the *ocs* locus of octopine T-DNA has been purified by hybrid selection (Murai and Kemp, 1982) (see Chapter 1). Because this RNA is more abundant in tumours than other T-DNA transcripts enough RNA was obtained for *in-vitro* translation analysis. The product whose synthesis *in vitro* it directed was a polypeptide of 39 000 molecular weight which was precipitated by antiserum raised against purified octopine synthase. Other polypeptides are synthesized by translation of T-DNA transcripts in wheat germ and rabbit reticulocyte lysate cell-free systems but have not yet been fully characterized.

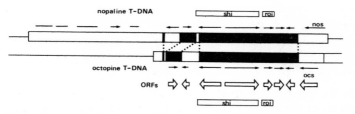

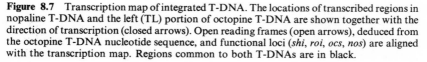

Figure 8.7 Transcription map of integrated T-DNA. The locations of transcribed regions in nopaline T-DNA and the left (TL) portion of octopine T-DNA are shown together with the direction of transcription (closed arrows). Open reading frames (open arrows), deduced from the octopine T-DNA nucleotide sequence, and functional loci (*shi, roi, ocs, nos*) are aligned with the transcription map. Regions common to both T-DNAs are in black.

8.10 The nucleotide sequence of octopine T-DNA

The complete nucleotide sequence of the T-DNA of the octopine Ti plasmid pTi 15955 has recently been determined (Barker *et al.*, 1983). In its 22 874-base-pair sequence resides the total genetic information controlling the neoplastic state. A computer search of the sequence has located the position of open reading frames delimited by protein synthesis initiation and termination codons. Open reading frames are potential protein-encoding genes and several of these align with the mapping positions of known RNA transcripts and functional genetic loci.

Octopine T-DNA integrates in two separate pieces, TL and TR. Transcription regulatory sequences adjacent to open reading frames within these two segments are typical of eukaryotic promoters and terminators (Chapter 3). The open reading frames are preceded by TATA boxes which are thought to specify the precise location of transcription initiation. Polyadenylation sites similar to the sequence AATAAA are found downstream of most open reading frames in the TL and TR regions. Thus, RNAs transcribed from these regions are probably regulated by host cell RNA polymerase II (see previous section). A short central region, TC, between TL and TR, in contrast exhibits features reminiscent of pro-karyotic DNA and its open reading frames do not have consensus eukaryotic promoter sequences.

The eight known transcripts of TL DNA map to open reading frames in this DNA segment (Figure 8.7). One identified transcript of TR DNA, which probably encodes the enzyme responsible for agropine synthesis, maps to one of the six open reading frames in the TR DNA portion. Taken together these observations suggest that integrated octopine T-DNA maintains the tumorous state of plant cells and specifies opine synthesis within them by expression of between 9 and 14 genes. It seems that none of the T-DNA genes are interrupted by introns. This is not a unique feature of eukaryotic genes in general or plant genes in particular, for example, the zein genes of maize lack intervening sequences as well (see Chapter 3).

8.11 Summary of the transformation process

A clearer picture is beginning to emerge of the molecular mechanisms involved in the neoplastic transformation of plant cells by *Agrobacterium*. Bacteria bind to wounded plant cells and transfer the Ti plasmid into a recipient plant cell by some as yet unknown process. Activity of virulence genes on the Ti plasmid control the integration of T-DNA into the plant genome. Following integration it is likely that the remainder of the inciting Ti plasmid is lost. Integrated T-DNA is then transcribed by RNA polymerase II and messenger RNAs translated on cytoplasmic ribosomes to produce T-DNA specific proteins. Some of these proteins maintain the tumorous state of plant cells and promote cell proliferation by influencing the phytohormone balance. The genes encoding these functions reside in the T-DNA common region. Opine synthesis is also directed by T-DNA gene products and opines secreted by tumour cells are catabolized by other bacteria, with a compatible Ti plasmid, around the site of the tumour. This confers a competitive advantage on the inciting *Agrobacterium* strain growing around the tumour since strains which do not contain in the correct Ti plasmid cannot utilize the particular opine produced by the tumour.

8.12 *Agrobacterium* and *Rhizobium*

Agrobacterium is very closely related to *Rhizobium* (see Chapter 7). Both organisms induce plasmid-directed alteration to plant growth. In the former case, the association is pathogenic to the plant, in the latter, it is beneficial. *Agrobacterium* Ti plasmid DNA integrates into host chromosomes; as far as is known, *Rhizobium* plasmid DNA does not, although this is still being investigated. In the evolutionary past, perhaps the two organisms diverged when they each acquired different types of plasmid DNA. It is possible that the T-DNA of the *Agrobacterium* Ti plasmid was originally captured from the host plant genome since it has features typical of eukaryotic DNA. Because many of the special characteristics of both *Agrobacterium* and *Rhizobium* are directly related to their plasmid complement, the distinction between strains and species of each organism is somewhat blurred, particularly in the light of the fact that plasmid transfer between strains can alter their properties in relation to the effects they have on plants.

Plasmid transfer between *Agrobacterium* and *Rhizobium* has been achieved experimentally. *Agrobacterium* harbouring a *Rhizobium* symbiotic plasmid can induce the formation of root nodules on plants. Following the reciprocal transfer, *Rhizobium* cells containing an *Agrobacterium* Ti plasmid induce a type of tumour formation on plants and these bacteria could also catabolize opines. However, nitrogen-fixing plant cells have not been produced as a result of plasmid cross-transfer between *Agrobacterium* and *Rhizobium*. Future work in producing nitrogen-fixing Ti plasmids could be agronomically beneficial because of the broad host-range of *Agrobacterium*. This would also necessitate the transfer of plant genes required in nodule formation, at present a formidable task (see Chapter 10).

CHAPTER NINE

PLANT VIRUSES

9.1 What are viruses?

Viruses are pathogenic agents and some of the diseases they cause have significant social and economic consequences for man. A knowledge of the mechanisms of virus multiplication at the molecular level can provide information about the disease syndrome and, perhaps, suggested the means for its control. Viruses are also useful tools for probing the molecular processes of cells. They have added greatly to our understanding of gene organization, expression and function in organisms of many taxonomic groups.

Viruses are genetic elements that subvert normal cellular processes to ensure their replication and maturation as transmissible entities. Virus propagation is specified by its own genome (either DNA or RNA) but always requires the energy generating and protein synthesizing machinery of host cells. The virus genome is usually protected, during its extracellular transmission phase, by encapsidation in a protein or lipoprotein coat. Viruses do not undergo binary fission; this is a characteristic of cells. They are non-cellular, obligate molecular parasites and some viruses require a second 'helper' virus to complete their replication.

Viruses of plants have features in common with those of other groups of organisms but there are important differences. Most plant viruses have relatively small genomes; none are known that attain the size of T-even bacteriophages or animal poxviruses. Plant viruses as a whole are unusual in that of the 300 or so characterized, greater than 90% have been shown to contain a genome of single-stranded RNA and in most of these, the RNA is of plus-(messenger) sense. Relatively few double-stranded RNA or DNA viruses of plants are known. Whether this is a true representation within the plant kingdom, or merely reflects an inability to detect or isolate viruses with certain genome types, remains to be seen. Another special feature is that the genomes of many plant viruses are subdivided into segments encapsidated in separate particles.

9.2 Biology of plant virus infections

Before virus infection of a healthy plant can occur, cell damage by disruption of the cell wall and/or plasmamembrane must precede virus entry. With invertebrate vectors of viruses, this is achieved during feeding. Mechanical inoculation, under experimental conditions, is effected by rubbing a virus suspension containing a suitable abrasive on to leaves of a susceptible plant. The source of infectious inoculum can be crude sap extracts from infected plants, preparations of purified virus and, for many different viruses, nucleic acid isolated from virions (the term for individual virus particles).

Following inoculation, symptoms usually appear on leaves at the sites of infection and these are known as local lesions. Some plants limit virus spread from the infection site by forming necrotic lesions (localized cell death) cutting off access to adjacent viable cells. This is known as the hypersensitive response. Plant resistance to virus infection is sometimes conferred genetically by one or more dominant or recessive host genes. There are mild strains of viruses which produce attenuated symptoms in plants. Pre-infection of a plant with a mild strain can confer cross-protection on the plant preventing super-infection by a second, more virulent, strain of the same virus. The mechanisms of cross-protection are not yet understood but presumably involve the expression of specific plant genes.

Following local infection, many plant viruses can spread systemically to other parts of the plant. Systemic symptoms of virus infection take on a variety of appearances from mosaic-mottling and yellowing of leaves to stunting of growth or plant death. Virus diseases in themselves do not always cause plant death but often debilitate plants so that they become susceptible to infection by other pathogens. However, some virus infections appear to have no obvious effects on the plant.

Little is known about the factors which govern the host-range of a particular virus. Alfalfa mosaic virus (AlMV) infects both monocotyledonous and dicotyledonous plants totalling at least 401 different species, whilst barley stripe mosaic virus (BSMV) is virtually restricted to barley. The complexities of host response, symptom expression and host-range will, however, be better understood when the functions of particular virus genes, and those of the host that are activated or repressed during infection, are elucidated.

On entry into a cell, the virus particle uncoats to release its nucleic acid. Translation of viral messenger RNAs to produce a number of polypeptides required in the replication cycle usually occurs on cytoplasmic 80S ribosomes. Synthesis of virus coat protein and genome replication is followed by packaging to produce progeny virus particles. Plant viruses induce a variety of cytological changes in infected cells, including abnormal membrane proliferation, changes in the appearance of organelles and accumulation of subcellular inclusion bodies or paracrystalline arrays of virus particles. The association of some viruses with nuclei, chloroplasts or other subcellular structures might indicate some functional significance for these organelles in virus replication.

An essential phase in the propagation of plant viruses is transmission to other plants. Many viruses are transmitted by invertebrate vectors such as nematodes, aphids or leafhoppers, which acquire virus during feeding. Some viruses, for example cauliflower mosaic virus (CaMV), induce the production of a 'helper' component in infected plants without which they cannot be acquired or transmitted by aphids. The helper component of CaMV is known to be encoded by a viral gene.

Experimental techniques used to study the biochemistry and molecular biology of cells are readily applicable to the analysis of viruses. Plant viruses can be isolated from infected tissues using methods similar to those for isolating cell organelles. Purified plant virus particles exhibit a variety of sizes and structural conformations that basically can be divided into three groups: (1) isometric (spherical), (2) bacilliform or bullet-shaped and (3) rods, which can be short, or long and flexuous. The particles of most plant viruses are nucleocapsids comprising an outer protein coat surrounding one or more nucleic acid molecules. Plant rhabdoviruses have, in addition, a lipid envelope.

A major contribution to our understanding of plant virus functions has come from *in-vitro* translation studies of plant viral messenger RNAs in cell-free systems derived from wheat-germ or rabbit reticulocyte lysates. However, *in-vitro* analysis must be complemented by studies of the virus replication cycle in plants. This is often difficult because infected plant tissues are composed of a heterogeneous population of cells producing virus asynchronously. Some of the problems of working with whole plants have been overcome by using protoplast systems. These wall-less cells can be infected *in vitro* by virus particles or viral RNA molecules and, in some systems, partial synchrony of infection and replication has been achieved, thus enabling an analysis of the time-course of events during the replication cycle.

9.3 Distribution of genome types among the plant viruses

Several schemes for classifying plant viruses have been devised, variously based on particle structure, serological relationships, host-range and genome composition. A convenient arrangement of the major groups of plant viruses, appropriate to our discussion in this chapter, has been described by Matthews (1982), based largely on genome organization (Table 9.1). Although many plant viruses have RNA genomes, the organization of genetic information differs between groups. The genome of most monopartite RNA viruses is a single-stranded RNA molecule of plus-sense (messenger polarity) and there are 11 major groups with this type of genome. Six major groups of plant RNA viruses have bipartite genomes. In these, the genome usually consists of two distinct plus-sense single-stranded RNA molecules that are encapsidated in separate particles. Both RNAs are required for infectivity in plants although examples are known where one of the two RNAs can support its own replication in protoplasts. Viruses with tripartite genomes (6 major groups) have 3 plus-sense single-stranded RNAs encapsidated in three separate particles. All three RNA molecules of tripartite viruses are required for infectivity in plants.

In addition to genomic RNA, many plant viruses have smaller sub-

Table 9.1 Plant virus genome types

	Genome type						
	ss[a] plus-sense RNA			ss minus-sense RNA	ds[b] RNA	ds DNA	ss DNA
	mono-partite	bi-partite	tri-partite				
No. of groups or families	11	6	6	1	1	1	1
Total no. of known members	152	46	24	9	10	7	13
No. of possible members	136	11	4	67	1	5	5

[a] single-stranded
[b] double-stranded

genomic messenger RNAs that are synthesized to amplify a specific gene product, for example coat protein. The nucleotide sequence of a subgenomic RNA is also present in one of the genomic RNA molecules and the former is derived from the latter, probably by a transcriptional mechanism involving the synthesis of a minus-strand RNA intermediate. The subgenomic RNAs of some viruses are encapsidated in particles together with one of the genomic RNAs whilst in other viruses they are encapsidated in separate smaller particles. Plant viruses often produce sub-genomic RNAs transiently during the replication cycle and these are not encapsidated in particles. Satellite viruses and satellite RNAs have been identified that associate with members of certain groups of plant viruses. The satellite virus (STNV) of tobacco necrosis virus (TNV) consists of a separately encapsidated messenger RNA which encodes the coat protein of STNV. Replication of STNV is entirely dependent upon the presence of the helper virus, TNV. The nucleotide sequence of satellite virus RNA shares no homology with that of the helper virus RNA. Satellite RNAs, for example CARNA 5 of cucumber mosaic virus (CMV), are distinct from satellite viruses. They are unrelated to, but entirely depend upon, the helper virus and become encapsidated in particles together with the genome of the helper. Satellite viruses and satellite RNAs influence the symptoms produced in plants by their respective helper viruses and can also interfere with helper virus replication and so can be considered as molecular parasites of the helper virus.

One group of plant viruses, the rhabdoviruses, share many characteristics of the similarly-named group of animal viruses. Their bacilliform, lipid-enveloped particles contain a single-stranded negative-sense RNA genome. Rhabdovirus messenger RNAs are synthesized from the genomic RNA template by a virion-associated transcriptase. Phytoreoviruses, again related to a group of animal viruses, have a complex genome of 10 or 12 segments of double-stranded RNA; they have the largest genomes of the known plant viruses.

Only two groups of plant DNA genome viruses are known: the caulimoviruses with double-stranded DNA and the geminiviruses. Some members of the latter groups contain a bipartite, whilst others a monopartite, single-stranded DNA genome.

9.4 Expression strategies of RNA virus genomes

Because the genomes of most plant RNA viruses are of plus-polarity, then the genome itself, or subgenomic fragments thereof, can be translated directly upon 80S ribosomes to produce virus-specific polypeptides. The basic anatomy of these RNAs is similar to host mRNAs with which they must compete for the protein-synthesizing machinery. This could simply be on the basis of numbers with viral mRNAs competing out the host mRNAs. Virus mRNAs can also exhibit a higher affinity for 40S ribosomal subunits enabling more efficient translation. *In-vitro* studies have shown that some viral mRNAs are translated more efficiently on 80S ribosomes than others, a factor that may be related to the length or composition of the 5' leader sequence.

The genome structure of plus-strand RNA viruses (Figure 9.1) is broadly similar between the different groups or families but there is variation in terms of modifications to RNA termini, translation and protein processing strategy. At the RNA 5' end three different types of structure have been identified. One is the cap ($m^7G^{5'}ppp^{5'}$), found on the RNA of at least 9 groups of plant viruses, a structure which also occurs at the 5' end of many eukaryotic mRNAs and is thought to enhance ribosome binding. A different modification is the VpG, a small protein of 3 000–7 000 molecular weight covalently linked to the RNA 5' end. Some animal picornaviruses (e.g. poliovirus) have a VpG and it has been suggested that this is involved in the initiation of RNA replication. Unmodified 5' termini with di- or tri-phosphate groups are present on yet other viral RNAs.

The 5' leader sequence of viral messenger RNAs precedes the first protein synthesis initiation codon (AUG). Ribosome 40S subunits bind to the leader and scan downstream for the AUG triplet before commencement of protein synthesis. The main body of a viral RNA contains protein-encoding nucleotides which start with AUG and end with one of three termination (stop) codons (UGA, UAG, UAA), delimiting an 'open reading frame' which is translated into a single polypeptide. Since eukaryotic ribosomes in general translate only that open reading frame closest to the RNA 5'-end (Kozak, 1983), those virus RNAs which have several internal protein encoding domains can generate sub-genomic

Figure 9.1 Generalized structure of a plus-strand RNA virus genome. An open reading frame is actually the nucleotide sequence between two in-frame protein synthesis termination codons but the term is often used to described a protein-coding cistron delimited by an initiation codon ($\triangledown$) and a termination codon ($\blacktriangledown$). The leader and UT are untranslated regions. The terminal modifications are discussed in the text.

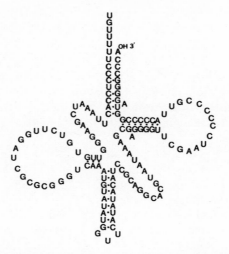

Figure 9.2   The secondary structure at the 3' end of TMV RNA. Predicted from the nucleotide sequence, it is reminiscent of tRNA and can also be aminoacylated. (Redrawn from Guilley *et al.*, 1979).

RNAs which bring the internal open reading frames to the 5'-end of smaller molecules. Processing of precursor 'polyproteins', translated from a single long open reading frame, into several smaller polypeptides, is an expression strategy adopted by several plant viruses. A 3'-end non-coding sequence followed by a poly(A) tract of 20–250 nucleotides is located on some plant viral RNAs in common with many host mRNAs. Viruses of at least 5 groups, however, are not 3'-polyadenylated but have an unusual tRNA-like structure (Figure 9.2) which can be aminoacylated with specific amino acids.

The variations in RNA plant virus genome structure and expression strategy (Hull and Davies, 1982) are illustrated in the following examples. The complete nucleotide sequences of tobacco mosaic virus (TMV), cowpea mosaic virus (CPMV) and brome mosaic virus (BMV) RNAs have been determined and current research on these viruses is directed towards an analysis of the functions of virus genes and gene products during the replication cycle in plants.

TMV

The monopartite genome of tobacco mosaic virus is a plus-strand RNA molecule of 6395 nucleotides (Goelet *et al.*, 1982)and the genomic RNA is encapsidated in rod-shaped particles 300nm long. Purified TMV RNA is infectious to plants. A cap structure is located at the 5′-terminus of TMV RNA; the 3′-end has a tRNA-like structure (Figure 9.2) that can be aminoacylated *in vitro* with histidine or valine (depending upon the virus strain). A 68-nucleotide leader sequence at the TMV RNA 5′-end precedes two long protein-coding open reading frames, taking up most of the length of the molecule (Figure 9.3). The first open reading frame is translated *in vitro* and *in vivo* to produce a polypeptide of moleculer weight 126 000. However, suppression of an amber termination codon at the end of the first open reading frame permits translation readthrough into the second, producing a larger polypeptide of 183 000 mol. wt. Both of these polypeptides appear in plants at relatively early stages of the infectivity cycle. Their functions have not yet been determined but it has been suggested that the 126 000 mol. wt. polypeptide contains the virus RNA replicase.

Two smaller open reading frames in TMV genomic RNA are located towards the 3′-end of the molecule. One encodes a 30 000 mol. wt. polypeptide, and the other the 18 000 mol. wt. coat protein. A defined RNA sequence within the 30 000 mol. wt. polypeptide cistron in TMV genomic RNA delimits the site at which coat protein attachment is initiated at the start of virion assembly. Sub-genomic RNAs are generated from the 3′-end

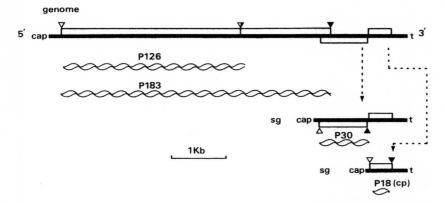

Figure 9.3 Genome strategy of tobacco mosaic virus. Genomic and sub-genomic (sg) RNAs have a 5′ cap and a 3′ tRNA-like structure (t). The open reading frame (open box) closest to the genomic RNA 5′ end is translated into a polypeptide of 126,000 mol. wt. (P126). Readthrough of a termination codon (▼) produces a larger polypeptide (P183). Subgenomic RNAs are derived from the 3′ end of the genomic RNA (dotted arrows) and one of these encodes the virus coat protein (cp).

sequences of TMV genomic RNA and one of these contains only the coat protein gene which is translated to produce the 18 000 mol wt. coat protein. Another, called I_2 RNA, contains both the 30 000 mol. wt. polypeptide and coat protein genes but only the former is translated because it is closest to the 5′ end of the sub-genomic RNA. It is interesting to note that I_2 RNA also contains the virion assembly initiation site and consequently becomes encapsidated in small rod-shaped particles (I-particles). The 30 000 mol. wt. polypeptide appears in plants early in the replication cycle and there is some tentative evidence to suggest it is involved in cell-to-cell spread of the virus.

CPMV

Cowpea mosaic virus has a bipartite genome, that is, its genetic material is coded on two plus-sense RNA molecules that are encapsidated in separate spherical particles. Three types of CPMV particles can be isolated from infected plants and then fractionated by sucrose gradient ultracentrifugation. The fast sedimenting bottom (B) component contains a single RNA species (B-RNA) 5889 nucleotides long (Lomonossoff and Shanks, 1983). An intermediate-sedimenting CPMV particle, the middle or M-component, contains the other genomic RNA (M-RNA) which is 3481 nucleotides long (Van Wezenbeek *et al.*, 1983). The slow-sedimenting 'top' component consists of empty virus capsids. The overall structure and organization of the two CPMV genomic RNAs is similar. A small protein, the VpG, is covalently linked to the 5′-end of each RNA and a poly(A) tail with an average length of 167 and 87 residues is present at the 3′-end of M-RNA and B-RNA respectively. A 5′ non-translated leader sequence of 160 nucleotides precedes a single long open reading frame of 3137 nucleotides in M-RNA whilst in B-RNA, the leader is 206 nucleotides long and its one open reading frame 5597 nucleotides long. The 3′-end non-translated region consists of 183 nucleotides in M-RNA and 85 nucleotides in B-RNA. There is a good deal of sequence homology in the 5′ and 3′ non-translated regions between the two RNAs. However, the long protein encoding open reading frame is unique to each molecule.

Both CPMV genomic RNAs are translated *in vitro* and *in vivo*, each producing a single polyprotein that is subsequently cleaved through a number of steps into several mature polypeptides (Figure 9.4). Functions have been ascribed to some of these proteins. Although both CPMV genomic RNAs are required for infectivity of plants, B-RNA can support its own replication, in the absence of M-RNA, following the *in-vitro*

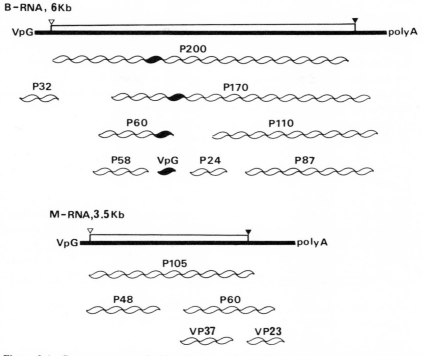

Figure 9.4 Genome strategy of a bipartite virus, CPMV. Each of the two genomic RNAs is translated to produce a polyprotein which is subsequently cleaved through a number of steps into smaller polypeptides. The virion coat proteins (VP37 and VP23) are derived from the M-RNA polyprotein and the VpG is coded by B-RNA.

inoculation of protoplasts. This demonstrates that a viral replicase is almost certainly coded by B-RNA. Because B-RNA cannot move to adjacent cells in plants in the absence of M-RNA, it has been suggested that M-RNA encodes a protein that mediates cell-to-cell spread. M-RNA also codes for the two virus capsid proteins of mol. wts. 23 000 and 37 000 (called VP23 and VP37) which are cleaved from the primary translation product. The small VpG, cleaved from the central portion of the B-RNA polyprotein, is thought to play a role in priming RNA replication.

A CPMV replication complex has been isolated from membrane fractions of infected cells and two of the B-RNA protein products may be important in anchoring this complex to plant membranes. Proteases required for specific processing of the M-RNA and B-RNA polyproteins are thought to be coded by B-RNA.

The genome strategy of CMPV bears a striking resemblance to that of animal picornaviruses such as poliovirus and foot and mouth disease virus. Each has a VpG and a poly-A tail and processes functional polypeptides from polyprotein precursors. The relative positions of specific polypeptide sequences (eg. capsid proteins, VpG etc) in the polyprotein are analogous in each virus. The major difference between the plant and animal picornaviruses, apart from the effects they have on their hosts, is that in the former group, the genome is divided into two molecules, whilst that of animal picornaviruses is a single RNA molecule.

BMV

The brome mosaic virus tripartite genome consists of three single-stranded plus-sense RNA molecules separately encapsidated in polyhedral particles 26nm in diameter. The sequences of BMV RNAs 1, 2 and 3 share little or no

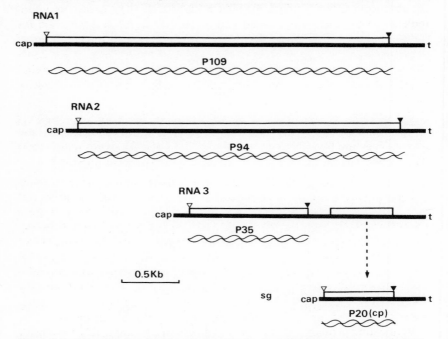

Figure 9.5 The tripartite genome of brome mosaic virus. The three BMV genomic RNAs each encode a single polypeptide. The second open reading frame of RNA 3, the coat protein gene, is expressed from a sub-genomic (sg) RNA which is also encapsidated in particles with RNA 3.

homology except for 200 nucleotides in the 3′ non-coding region common to each molecule. The genomic RNAs have a 5′ cap and a tRNA-like structure at the 3′-end that can be aminoacylated with tyrosine. BMV RNA 1 (3234 nucleotides) and RNA 2 (2865 nucleotides) (Ahlquist *et al.* 1984) both contain a single, long open reading frame translated to produce polypeptides of mol. wts. 109 000 and 94 000 respectively. RNA 3 (2117 nucleotides) has two open reading frames but only the 5′-proximal one is translated, into a 35 000 mol. wt. polypeptide. The second open reading frame of RNA3 is expressed from a sub-genomic mRNA (BMV RNA 4) derived from the last 876 nucleotides of RNA 3. RNA 4 encodes the 20 000 mol. wt. BMV capsid protein. RNAs 3 and 4 are encapsidated together in one particle but only RNAs 1, 2 and 3 are required for infectivity and thus constitute the total genetic information of the virus genome (Figure 9.5).

Relatively little is known about the functions of BMV polypeptides. Barley protoplasts support replication and translation of RNAs 1 and 2, in the absence of RNA 3, and so it is likely that one of them codes for a viral replicase. The function of the translation product of RNA 3 is not known but by analogy with TMV and CPMV it could specify cell-to-cell spread of the virus. It has been proposed that the homologous sequences at the BMV RNA 3′-ends regulate gene expression during the replication cycle and that the tRNA-like structure primes RNA synthesis by some, as yet unknown, mechanism.

The genome strategies of the foregoing plant RNA viruses illustrate the types of mechanisms that are adopted to achieve virus propagation within plants. They are by no means the only genome strategies but many other plant RNA viruses exhibit variations around these general themes.

9.5 RNA virus replication

Relatively little is known about the mechanisms of RNA virus replication. Most studies have involved isolating double-stranded RNAs composed of plus-strands base-paired to minus-strands. Replicative forms (RFs) consist of full-length double-stranded RNAs, whereas replicative intermediates (RIs) are double-stranded molecules with one strand full-length and the other strand growing and consequently of varying length producing molecules that are partially single- and partially double-stranded.

Replication probably starts close to one end of a genomic RNA molecule primed by a specific terminal structure. Transcription of plus-strands into minus-strands and *vice versa* may be directed by a virus-coded replicase. However, many plants, not infected by viruses, contain RNA-dependent

RNA polymerase activity and, in some virus/host combinations at least, virus replication could be controlled by a host-coded enzyme.

9.6 DNA viruses

CaMV

Two groups of plant DNA viruses have been identified. Members of one of these, the caulimoviruses, have a double-stranded DNA genome and the type-member of the group, cauliflower mosaic virus (CaMV), has been extensively studied as a potential plant gene cloning vector (see Chapter 10), and consequently is one of the best characterized plant viruses

(a)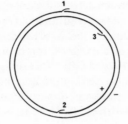

(b)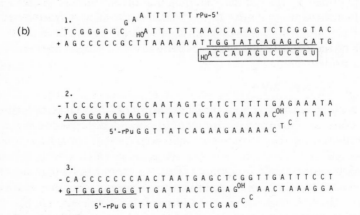

Figure 9.6 The discontinuities in cauliflower mosaic virus DNA. (*a*) Diagram of the circular ds DNA genome of CaMV showing the location of gaps (1–3) in each strand. (*b*) Nucleotide sequence at each discontinuity with the regions thought to be involved in priming DNA synthesis underlined. Boxed sequence is part of the plant tRNA^{met} primer.

at the molecular level (Hohn *et al.*, 1982). Each of the 12 or so members of the caulimovirus group has a fairly narrow host range; CaMV, for instance, infects only brassicas. Caulimoviruses have isometric particles about 50nm in diameter and the cells of infected plants accumulate dense, virion-containing, proteinaceous inclusion bodies thought to be the sites of virion assembly.

DNA extracted from caulimovirus particles is infectious to plants and consists of circular double-stranded molecules of about 8000 nucleotide pairs. An unusual feature of the DNA is that single-stranded discontinuities or gaps are located at specific sites on each strand. One DNA strand always has only one gap whilst the second strand contains 1, 2 or 3 gaps depending upon the virus or virus strain. Most strains of CaMV have 2 gaps in the second DNA strand. The gaps consist of short regions of triple-stranded DNA; the 5′ end of each strand is at a fixed position but the 3′ end varies in length from 5–40 nucleotides producing an overlap or flap close to the gap (Figure 9.6). Ribonucleotides are associated with the gaps and it is believed that they represent origins of DNA replication (Guilley *et al.*, 1983).

Nucleotide sequencing of CaMV DNA (Franck *et al.*, 1980) has revealed 6 long open reading frames on only one strand, the minus strand, and so genetic information is expressed asymmetrically by transcription of the minus-strand into mRNA (Figure 9.7). In virion DNA, the minus-strand has a single gap but the template for transcription is a covalently-closed circular (gapless) form of CaMV DNA found in the nuclei of infected cells which arises following repair of the virion DNA gaps (Olszewski *et al.*, 1982).

Four of the open reading frames of CaMV DNA have been assigned specific protein coding functions. The 57 000 mol. wt. capsid protein precursor (subsequently cleaved to a mature 43 000 mol. wt. polypeptide that is phosphorylated and glycosylated) is a product of gene IV. The major constituent of subcellular inclusion bodies, found in infected plants, is a 62 000 mol. wt. polypeptide product of gene VI. An 18 000 mol. wt. polypeptide encoded by gene II is somehow involved in the acquisition and transmission of CaMV by its aphid vector. Since this later function is dispensable to the virus replication cycle within plants, it is possible to delete gene II by *in-vitro* mutagenesis without interrupting essential virus functions. Such deletions in gene II, however, result in virus particles that cannot then be acquired and transmitted by aphids. Gene V is thought to encode a polymerase involved in the rather unusual mode of replication exhibited by CaMV.

Each gene in the virus DNA must be expressed by transcription into

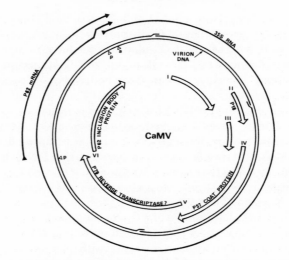

Figure 9.7 Genes and RNA transcripts of CaMV. The CaMV genes (I–VI) are located on only one DNA strand. Two polyadenylated RNA transcripts have been identified; one is the mRNA for gene VI, the other (35S RNA) has a short terminal repeat sequence. Transcription promoters (p) and a poly-A signal sequence (a) are shown.

translatable mRNAs. Two major CaMV RNA transcripts have been characterized in detail. The mRNA expressing gene VI is a capped, polyadenylated molecule 1850 nucleotides long (Figure 9.7). The second major CaMV transcript, called 35S RNA, is also polyadenylated. Since 35S RNA is transcribed from one complete DNA strand encompassing all of the open regions, it might represent a possible precursor to smaller mRNAs. However, subgenomic mRNA transcripts expressing CaMV genes (with the exception of gene VI) have not been characterized probably because they are present in low concentrations in infected cells. A second possibility, that 35S RNA is translated as a polycistronic messenger, at present, seems unlikely because eukaryotic ribosomes, in general, translate only that open reading frame closest to the 5′-end of an RNA molecule. Several, unanswered questions concerning the mode of CaMV gene expression await further research.

Transcription of both gene VI and 35S RNAs appears to be separately regulated by CaMV DNA sequences upstream of the 5′-end of each transcribed region that have typical characteristics of eukaryotic promoters recognized by RNA polymerase II. An unusual feature of 35S RNA is that its transcription terminates 180 nucleotides downstream of the site of its transcription initiation after one complete circuit of the DNA. This produces a molecule with a terminal directly-repeated sequence of 180 nucleotides (Figure 9.7).

Until fairly recently little was known about the mechanisms of CaMV DNA replication; however, a number of observations have revealed certain striking similarities between CaMV and a group of unusual animal RNA tumour viruses, the retroviruses, which replicate by reverse transcription of their RNA genome into DNA, then transcribed back into RNA. These similarities in CaMV include the coding of protein on only one strand, the existence of a genome-length RNA transcript (35S RNA) with terminal-repeated sequences and close homology of nucleotide sequences near the origins of plus- and minus-strand DNA synthesis (the CaMV DNA gaps) (Figure 9.6).

The CaMV replication cycle as it is currently understood (see Hull and Covey, 1983) can be summarized as follows (Figure 9.8). Virion DNA molecules, containing gaps, enter the plant cell and move to the nucleus where the gaps are sealed and a supercoiled molecule is generated that associates with host proteins in a minichromosome structure. The minus-strand of CaMV DNA in the minichromosome is transcribed by host RNA polymerase II into mRNAs that move to the cytoplasm for translation into protein. The genome length 35S RNA transcript is also synthesized in the nucleus, although it is not certain whether replication of CaMV occurs in the nucleus or cytoplasm. The proposed model for CaMV replication is directly analogous to reverse transcription of animal retroviruses. Synthesis of CaMV minus-strand DNA is probably primed by a host cell tRNA^{met} which hybridizes to a 14-nucleotide sequence in 35S RNA (known to be complementary to tRNA^{met}), located some 600 nucleotides downstream from the 5'-end of 35S RNA and adjacent to one of the CaMV DNA gap sequences. Reverse transcription proceeds to the 5'-end of 35S RNA to produce a small minus-strand CaMV DNA fragment (called strong-stop DNA) still covalently linked to the tRNA primer (Turner and Covey, 1984). Because 35S RNA has a direct terminal repeat sequence, strong-stop DNA can circularize the RNA template and reverse transcription continues to complete the DNA minus-strand.

Plus-strand DNA synthesis is primed by G-rich sequences located near to the gaps found in virion DNA (Figure 9.6); the completed DNA minus-strand is the template for plus-strand synthesis. At the end of the replication cycle, the strand overlaps at each gap in CaMV virion DNA, result from limited strand displacement synthesis by the reverse transcriptase. Many of the replication intermediates predicted by this model

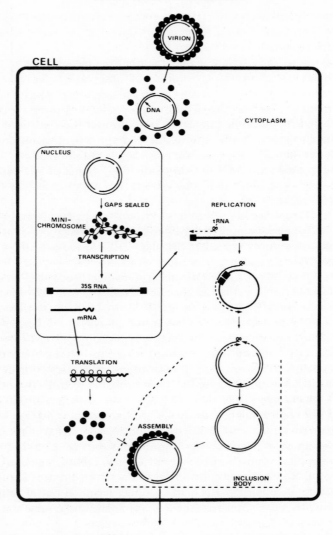

Figure 9.8 The replication cycle of CaMV. See text for discussion.

have been detected in CaMV-infected plants but the purification and characterization of the CaMV reverse transcriptase enzyme is still eagerly awaited. The idea that reverse transcription occurs in plants is, at first, quite surprising. Perhaps a more detailed study of plant transposable elements (see Chapter 2) will reveal even greater surprises because one group of these elements found in yeast, *Drosophila* and vertebrates also shares structural similarities with retroviruses and CaMV.

Geminiviruses

Geminiviruses are the only other group of plant DNA viruses known. The group name originates from the appearance of purified virus particles under the electron microscope: they are twin spheroids. Some thirteen different geminiviruses have been identified and they infect several families of monocotyledonous and dicotyledonous plants. About half of the known geminiviruses are transmitted from plant to plant only by whitefly vectors; the other half only by leafhoppers (see Davies and Hull, 1983). A characteristic feature of cells infected by some geminiviruses is the presence of doughnut-shaped or paracrystalline inclusions in the nucleus. This suggests that the nucleus is important in at least some steps in the virus replication cycle.

DNA isolated from geminiviruses, such as bean golden mosaic (BGMV) and cassava latent (CLV) viruses, consists largely of single-stranded open circles. Restriction enzyme mapping and nucleotide sequencing of the DNA has shown that there are two distinct circular DNA types with different nucleotide sequences. Surprisingly, each geminate (paired) particle contains only one of the two DNAs and thus, two distinct types of geminate particles exist. These viruses, therefore, have a bipartite genome. DNA 1 of CLV consists of 2780 nucleotides and DNA 2 2724 nucleotides. Although the nucleotide sequences of DNAs 1 and 2 are mostly different, a region of 219 nucleotides is common to both (Stanley and Gay, 1983).

Studies of geminivirus replication and transcription are in their early stages. Double-stranded forms of geminivirus DNA have been detected and are likely replication intermediates. The common region of CLV DNAs 1 and 2 contains a nucleotide sequence reminiscent of a DNA replication origin in that it has a hairpin loop-structure. Several long open reading frames, potential protein encoding genes, are located on both DNAs of CLV. The coding functions of these are not yet known, with the exception of one open reading frame on DNA 1, which is the coat protein gene.

Maize streak virus (MSV) and chloris striate mosaic virus (CSMV) each appear to have only one single-stranded DNA circle. Why these differ from CLV, BGMV and others is not yet clear. A rather strange observation is that whitefly-transmitted geminiviruses have two different DNA molecules comprising their genome, and are thus bipartite, but leafhopper transmitted geminiviruses seem to have only one. This needs further investigation as only relatively few geminiviruses have been studied in detail (Goodman, 1981).

9.7 Viroids and virusoids

Viroids are most intriguing entities. They produce disease symptoms in plants similar to those of viruses. Early attempts to isolate viruses from some plants with certain virus-like diseases were unsuccessful. Instead, diseased plants were eventually found to contain very small RNA molecules that were absent from plants not showing symptoms. Infectivity studies using preparations of these small RNAs demonstrated that they were the causative agent of the disease.

Viroids, as they became known, are RNA molecules 200–400 nucleotides in size (Sänger, 1982) and they differ from viruses in that they are naked, that is, not encapsidated in coat protein. Viroids are the smallest known infectious entities and strictly speaking they cannot be described as viruses. Electron microscope analysis of purified potato spindle tuber viroid (PSTV) has shown that the viroid RNA appears to consist of double-stranded rods about 50 mm long. However, following denaturation, single-stranded open circular molecules were observed. Thus, viroid RNAs are single-stranded circular molecules with a high degree of intramolecular base-pairing, producing a rod-like configuration. Several viroid RNAs have been sequenced: PSTV consists of 359 nucleotides (Figure 9.9) and citrus exocortis viroid (CEV), 371 nucleotides. Certain regions of the RNA of different viroids exhibit a high degree of sequence homology.

How do viroids function in plant cells? It is now generally accepted that viroid RNA does not encode protein. No viroid-specific protein has been detected in infected plants, viroid RNA is not translated *in vitro* and there are no long open reading frames in the viroid RNA sequence which could code for proteins of appreciable size. One important experimental advance has been the finding that *in-vitro* synthesized full-length complementary DNA copies of viroid RNA, after cloning in *E. coli*, are infectious to plants which then produce progeny viroid RNA (see Robertson *et al.*, 1983). Site-directed mutagenesis of cDNA clones of viroid RNA should help in identifying functional sequences. However, two naturally-occurring strains of PSTV have been isolated that cause either mild or severe symptoms in plants. The nucleotide sequences of these two strains differ by only 3

Figure 9.9 Nucleotide sequence and secondary structure of potato spindle tuber viroid. The arrows denote the positions of base changes which convert the severe strain into the mild strain.

nucleotides out of 359! It seems that in terms of symptom expression, the viroid RNA sequence has a very precise function indeed.

Two speculative ideas that attempt to account for the molecular pathogenicity and origin of viroids have been suggested. The processing of precursor messenger RNA transcribed from split genes, by splicing, is thought to be mediated by base-pairing interaction with a small molecule called U1 RNA (see Chapter 3). Viroids have sequence homology with U1 RNA and could interfere with the splicing mechanism. Certain regions of viroid RNA also appear to be remnants of sequences found in one group of eukaryotic transposable elements with structures similar to animal retrovirus DNA. It has been suggested, because of these similarities, that viroids 'escaped' from transposable elements in the evolutionary past (Kiefer et al., 1983). There is, however, no evidence that viroids replicate by synthesizing a DNA intermediate.

Virusoids are viroid-like RNA molecules found encapsidated with the genomic RNA of certain plant viruses (see Robertson et al., 1983). Although virusoid RNA is structurally similar to viroid RNA, it is incapable of independent replication and relies for this on its helper virus. The virusoid associated with lucerne transient streak virus (LTSV) shares no sequence homology with the genome of its helper virus. LTSV can, however, replicate independently of the virusoid RNA. Some other viruses, for example velvet mottle virus, appear, in contrast, to be entirely dependent on the presence of virusoid RNA without which the virus cannot replicate.

CHAPTER TEN

PROSPECTS FOR THE GENETIC ENGINEERING
OF PLANTS

Over the past 50 years or so conventional plant breeding, combined with improved agricultural practices and modern technology, has contributed to a dramatic increase in the production of plants for food. Not all countries have achieved the same increases in food production, however, and many are unable to feed their own populations. The high agricultural production in some developed countries depends upon favourable climatic conditions, high input of fertilizers and crop protection chemicals, and the successful development of new varieties. It is doubtful whether poorer countries will be able to, or even wish to, develop this type of agriculture. Even in countries with a modern food production industry, the input costs are very substantial. Furthermore, there are limitations to the geographical distribution and quality of certain crops. There is, therefore, a continuing and dual pressure to grow more food at less cost and to improve existing crops in order to make them more suitable for our needs.

10.1 Modern plant breeding

Recently there has been a great deal of interest in the possibility of using the recombinant DNA techniques to achieve some of the objectives listed in Table 10.1. However, it is important to recognize that many improvements can be accomplished by conventional plant breeding; there is still tremendous scope for utilizing the wide range of existing germ plasm to transfer cold-tolerance, disease-resistance or other useful qualities to food plants. The design of a selection system for characteristics such as these is relatively simple but selection for other desirable attributes is not always so straightforward. Molecular biology can provide information about the molecular basis for specific plant characters which can aid in the selection process. An example of this is the identification of lines of wheat which make good leavened bread. The old test involved accumulating sufficient

Table 10.1   Some possible goals for plant improvement

1. Reduce or abolish photorespiration in C3 plants, thus increasing net carbon fixation.
2. Extend the capacity to fix atmospheric nitrogen to cereals or other major crops.
3. Develop plant varieties that are tolerant to high salinity or flooding.
4. Improve water economy of plants and develop drought-resistant varieties.
5. Generate herbicide-resistant crop plants that are unaffected by sprays that kill competing weeds.
6. Improve the resistance of plants to diseases.
7. Modify the amino acid composition of storage protein in cereals and legumes to improve their nutritional value.
8. Introduce cold-tolerance into plants of tropical and sub-tropical origin.
9. Improve the composition and storage life of fruits and vegetables.

grains to make a loaf; samples of the material that produced good loaves were used subsequently by the breeders. They now analyse the storage proteins in the grains by gel electrophoresis, for example, to study the glutens which make a major contribution to bread-making quality. This provides a simpler, more objective test for screening.

Other new methods that can contribute to plant breeding have become available through studies on tissue culture. For example, breeders can exploit the somaclonal variation that occurs when plants are regenerated from tissue culture and select for new improved lines. Furthermore, the successful fusion of protoplasts from different species and the regeneration of viable hybrids opens up the possibility of developing agriculturally useful plants from novel unions.

Although these methods may prove to be extremely useful, they depend ultimately upon chance and selection and do not involve the identification and manipulation of genetic targets which are defined in the molecular sense. The potential value of genetic engineering is in providing a major new approach capable of achieving objectives not possible by other means. The general strategy would be (1) to identify and isolate DNA sequences which control processes important in plant growth and productivity and (2) to modify the existing genes or transfer new ones from other organisms, so that the performance of a particular plant is improved. Both of these approaches are being actively investigated and much effort has been directed towards identifying and developing systems capable of delivering genes, in the form of specific DNA sequences, to recipient plants.

10.2 Gene vectors

In the context of plant genetic engineering, the term 'vector' is used to describe a biological carrier that takes DNA across the cell wall barrier into a recipient cell where it is expressed. Likely candidates for vectors are those biological systems where entry of nucleic acid into a plant normally occurs pathogenically. This happens with *Agrobacterium* Ti plasmids and with plant viruses, and these are being studied with the aim of developing controllable gene vectors. The idea is that 'foreign' genes are hitched onto the vector and carried with it into the plant.

Agrobacterium *Ti plasmids*

As we have seen (Chapter 8), the T-DNA portion of the *Agrobacterium* oncogenic Ti plasmid is transferred to the nucleus of a host plant where it integrates into the genome and is then transcribed to produce mRNAs. The Ti plasmid is therefore a natural gene vector, but can it accommodate extra DNA sequences without affecting its integration or functions? Functional loci have been mapped by inactivation of specific T-DNA genes following insertion of bacterial transposons. In one such experiment, Hernalsteens *et al.* (1980) inserted the transposon Tn7 into the nopaline synthase (*nos*) gene and Tn7 DNA sequences were transferred with the T-DNA to the recipient plant. Although the resulting tumour cells failed to synthesize nopaline, foreign DNA (Tn7) had been stably integrated with the T-DNA into the plant genome. Subsequent experiments have shown that at least 50 kilobase pairs of foreign DNA can be integrated into the nuclear genome mediated by T-DNA transfer; the upper size limit of inserted DNA is yet to be determined.

Bacterial transposons carry genes encoding antibiotic resistance that are normally expressed in bacteria. It was found in early experiments that these genes were not expressed when integrated with T-DNA. This is because the prokaryotic regulatory sequences controlling transposon gene expression were not recognized by plant enzymes transcribing T-DNA. Clearly, in order to get expression of a foreign gene, it would seem necessary that the appropriate plant transcription signals (promoters, terminators etc.; see Chapter 3) are located in the correct positions adjacent to the inserted DNA. To test this, Herrera-Estrella *et al.* (1983) deleted the *nos* gene in a nopaline Ti plasmid pTiC58 and inserted in its place an octopine synthase (*ocs*) gene obtained from an octopine Ti plasmid. The *nos* promotor was left intact upstream of the newly-inserted *ocs* DNA sequence. Following integration into the plant genome, the resulting tumour cells synthesized octopine under direction of the *nos* promoter. In similar construct experiments, a bacterial gene encoding chloramphenicol acetyl transferase was expressed in plant tumours under direction of a *nos* promoter.

Transfer of foreign plant genes has also been achieved using the Ti plasmid as a vector. Murai *et al.* (1983) have inserted the gene encoding a bean (*Phaseolus vulgaris*) storage protein, phaseolin, into an octopine Ti plasmid pTi15955. This plasmid construct was then used to transform sunflower plants. The resulting tumour cells were found to contain RNA transcripts of the phaseolin gene and phaseolin polypeptides. There are two important factors about these experiments: first, the integrated phaseolin gene contained five intervening sequences (introns) and these were correctly excised from the primary RNA transcript by splicing; T-DNA gene do not contain introns, so far as it is known. Second, expression of the phaseolin gene in sunflower tumour cells was more efficient when under direction of the T-DNA *ocs* promoter than under its own promoter. This is interesting because phaseolin synthesis in bean plants is developmentally regulated and the gene is activated only before seed storage protein is laid down after flowering. DNA sequences that control differential gene expression will eventually be required as components of a plant gene vector to ensure that inserted genes are active only at particular stages of a developmental sequence.

Regeneration of tumour cells into whole, fertile plants is an important requirement following Ti plasmid-mediated genetic engineering. *Agrobacterium*-induced plant tumours usually cannot be regenerated into normal plants that can transmit the T-DNA through meiosis to progeny plants. However, Ti plasmid deletion mutants have been isolated that have lost most of the central region of the T-DNA encoding oncogenic determinants, but not opine synthase genes. Plant cells transformed by such mutants retain the ability to synthesize opines and apparently normal plants have been regenerated from them which still make opines; a characteristic subsequently passed on to progeny plants.

Disarming the oncogenic properties of Ti plasmids whilst still maintaining the ability to integrate DNA containing selectable characteristics (opine synthesis, antibiotic resistance etc.) is another essential consideration in constructing a gene vector.

Ti plasmids are very large molecules and the insertion and deletion of specific DNA sequences involves complex *in-vitro* and *in-vivo* recombination techniques. Smaller molecules are easier to handle experimentally than the very large Ti plasmids. Current ideas in developing Ti plasmid-based gene vectors centre around constructing mini-Ti vectors which have most of the T-DNA deleted. The Ti plasmid virulence (*vir*) region still directs the insertion of T-DNA into the plant genome when it is located on a separate plasmid molecule (de Framond *et al.*, 1983). Since only the T-

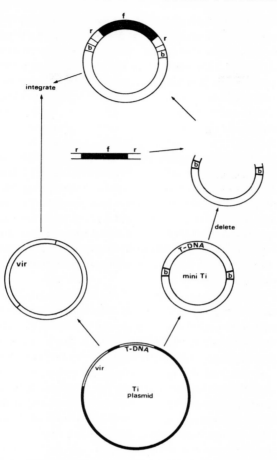

Figure 10.1   The mini Ti plasmid gene vector. The virulence (*vir*) region and T-DNA of the Ti plasmid are first separated on to two smaller plasmid molecules. The T-DNA is then deleted except for its border sequences (b). A foreign gene (f) also encoding a selectable marker, flanked by plant transcription regulatory sequences (r), is inserted between the T-DNA borders. The *vir* plasmid mediates integration of the border fragments and the foreign gene into the plant genome where it is transcribed under direction of the flanking regulatory sequences.

DNA 24-base-pair border and adjacent sequences appear to be required for integration (Caplan *et al.*, 1983), a second small plasmid containing these sequences plus a selectable inserted gene, with the appropriate transcription control elements, would be a more readily manipulable vector with the integration mechanism directed by *vir* genes located on a different molecule (Figure 10.1).

Plant DNA viruses

At first sight, the double-stranded DNA genome of cauliflower mosaic virus (CaMV) (see Chapter 9) has several attributes of a plant gene vector: the purified viral DNA is infectious when rubbed on to the leaves of susceptible plants. The DNA is also infectious after it has been cloned into *E. coli* plasmids (Howell *et al.*, 1980) and is therefore amenable to various *in vitro* manipulations. After infection, CaMV spreads to virtually every cell in the host plant: infected cells contain about 10^5 copies of the virus genome and this is matched by an appreciable accumulation of virus gene products. However, CaMV has certain disadvantages as a vector compared with the *Agrobacterium* Ti plasmid. The host-range of CaMV and other caulimoviruses is relatively limited. It may be possible to overcome this limitation when more is known about factors specifying host-range. Unlike *Agrobacterium* T-DNA, CaMV DNA appears not to integrate into host chromosomes. This may not be such a disadvantage because plants can be propagated vegetatively, thus obviating the requirement for meiosis in transmitting engineered genes to progeny plants (Hull, 1978). Deletion and insertion mutagenesis experiments designed to locate non-essential regions in CaMV DNA that might accommodate foreign genes have produced some rather disappointing results. Most deletions render CaMV DNA non-infectious and those few sites that can be deleted without affecting infectivity will not accommodate additional foreign sequences larger than 200–300 nucleotides (Dixon *et al.*, 1983; Daubert *et al.*, 1983). This may be because of packaging constraints during DNA encapsidation or interference in the way CaMV genes are normally expressed in plant cells. Attempts to complement defective or deleted CaMV genomes have also proved problematical because recombination in the plant produces normal genomes from pairs of defective but complementary CaMV genome fragments (Walden and Howell, 1982). This high rate of recombination exhibited by CaMV probably reflects its rather unusual mode of replication, apparently involving reverse transcription.

If the CaMV genome as a whole does not seem to be a very promising plant gene vector at present, its contribution may come from some of its sequences. CaMV has two 'strong' transcription promotors which might be useful components of a chimaeric gene vector. Studies of CaMV could also provide information about the processes that regulate gene expression in plants, an area currently requiring intensive investigation.

Geminiviruses are the other group of plant DNA viruses attracting attention as possible gene vectors. Relatively little is known about the biology of these viruses, although the single-stranded DNA genome of one member of the group (cassava latent virus) has been sequenced and the DNA cloned into bacterial plasmids is infectious to plants (Stanley, 1983). Maize streak virus is a geminivirus of particular interest since it infects agronomically important monocotyledonous plants. *Agrobacterium tumefaciens* does not infect this group of plants and the difficulties in manipulating cell cultures of monocotyledons makes a disabled virus vector an attractive candidate for genetic manipulation. It is not yet known whether geminivirus DNA has non-essential regions which might accommodate foreign DNA sequences.

Plant RNA viruses

It is probable that all higher plants are hosts to several types of RNA genome viruses, although many of them have not to date been thoroughly characterized. Thus, the difficulties in overcoming host-range limitations exhibited by the previously-mentioned pathogen-based vector systems could be circumvented by choosing an RNA virus with the appropriate host-range for modification to carry foreign genes into any desired plant (Hull and Davies, 1983). This sounds fine in principle but RNA molecules are, at present, less readily manipulable *in vitro* (cut, joined, cloned etc.) than DNA molecules, and constructing vectors from them presents several technical problems. However, it is possible to generate complementary DNA copies (cDNAs) of RNA molecules *in vitro* using the enzyme reverse transcriptase (Chapter 1). There is evidence that DNA copies of RNA virus genomes retain infectivity. Full-length double-stranded cDNA copies of a bacterial RNA virus ($Q\beta$) (Taniguchi *et al.*, 1978) and an animal RNA virus (poliovirus) (Racaniello and Baltimore, 1981) were found to be infectious when introduced into their respective host organisms. This has not yet been demonstrated for plant RNA viruses (with the exception of viroids—Chapter 9) but remains an exciting possibility (Figure 10.2).

Encapsidation of foreign nucleic acid in the coat protein of rod-shaped RNA viruses (which do not have the limited packaging constraints of

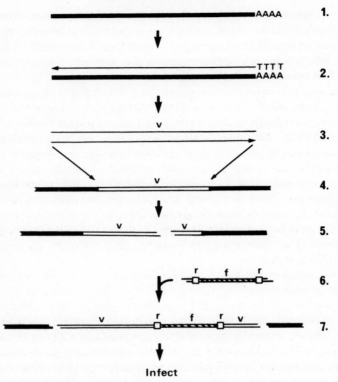

Figure 10.2 Speculative scheme for developing a gene vector from plant RNA genome viruses. 1. Polyadenylated RNA virus genome converted to a double-stranded cDNA *in vitro* (2, 3). 4, the virus cDNA (v) is cloned into a bacterial plasmid for easy manipulation and digested with an appropriate restriction enzyme (5). 6, a foreign gene (f) with flanking transcription signals (r) is inserted into the virus cDNA. The chimeric molecule is excised from the bacterial plasmid (7) and used to infect plants.

spherical viruses) is another possible means of introducing foreign genes into plant cells.

Transposable elements

Plant transposable elements have been most thoroughly studied in maize (see Chapter 2) but there is still much to learn about their structure and activity at the molecular level. By analogy with other eukaryotes, plants are likely to contain a variety of DNA sequences with characteristics of transposable elements. Their potential as genetic manipulation vectors is based upon apparent ability to integrate at multiple sites within the nuclear genome. The prospects for using transposable elements as gene vectors are not without precedent in eukaryotes. Foreign genes have been integrated into the nuclear genome of *Drosophila* mediated by a transposable element. The inserted genes were subsequently expressed under developmental regulation. Transposon-mediated integration still requires that foreign DNA is initially inserted into the recipient cells. In *Drosophila*, this was achieved by microinjection of DNA into eggs. Although it may be possible to construct chimaeric gene vectors with DNA sequences derived from the *Agrobacterium* Ti plasmid, CaMV or some other naturally infectious agent, other more direct methods of inserting DNA into plant cells are being investigated as well.

Direct transformation of plant cells

In several animal systems, it is possible to insert directly DNA molecules into cells by incubating them with DNA-containing solutions under the appropriate conditions. Calcium phosphate is usually used to mediate the transfer. Attempts to transform plant protoplasts directly with DNA have generally been unsuccessful, although it has been done with *Agrobacterium* Ti plasmid DNA. RNA isolated from some plant viruses has been used to inoculate protoplasts *in vitro* that subsequently produced progeny virus particles. So it is possible to insert directly nucleic acid into individual plant cells where it is replicated and expressed. Microinjection of DNA into cells also works well in many animal systems and *Xenopus* oocytes are the best example of this. These cells are very large indeed (up to 1 mm in diameter) so microinjection of DNA into them is a relatively straightforward procedure. Most plant cells are much smaller and with many types of differentiated plant cell, 90% or more of their volume is taken up by a large vacuole. One rather elegant way round this problem is to microinject DNA directly into the nucleus or cytoplasm of germinating pollen tubes.

10.3 Identifying genes

How justified are we in believing that genetic engineering can be used successfully to modify plants? What types of improvements should we be aiming for, and how can we begin to go about achieving them? The following are a few examples of particular characteristics of plants which might be amenable for manipulation by molecular means.

Herbicide resistance

If crop plants were resistant to herbicides, this would lead to more effective weed control, which could increase yields. Such a possibility is being investigated in relation to the triazine group of herbicides which kill plants by interfering with photosynthetic electron transport. The photoaffinity label azido-atrazine has been used to locate the herbicide binding site in the chloroplast thylakoids. It is believed to be a 32 000 molecular weight protein (the product of photogene 32; see Chapter 4), and this is encoded by the plastome. More than twenty weed species have developed resistance to triazine herbicides. In one case, this has been correlated with a single base change in photogene 32, resulting in a change from serine to glycine in the protein. Conventional breeding between the resistant weed *Brassica campestris* and *Brassica napus* (oilseed rape) has led to the production of resistant crop plants (Beversdorf *et al.*, 1980). Experiments are also being carried out in an attempt to transfer resistance from *Solanum nigrum* (black nightshade) to *Solanum tuberosum* (potato), using protoplast fusion. The real challenge, however, is to try and achieve the appropriate change in crop plants for which there is no close resistant relative.

Improving photosynthetic efficiency

The efficiency of photosynthesis is increased when carbon dioxide concentration is high and oxygen is low, thus favouring the carboxylation reaction of rubisco over the oxygenase reaction (Chapter 4). This occurs in C4 plants which can maintain a high CO_2 concentration in the region of rubisco, but not in C3 plants, which lose an estimated 25% of fixed carbon by photorespiration. Over five million mutants have been screened in an unsuccessful attempt to generate *Arabidopsis* plants deficient in the photorespiratory pathway. It has been concluded that the only possible solution to the problem is to attempt to reduce or abolish the oxygenase activity of rubisco (Somerville and Ogren, 1982), which is located on the large subunit encoded by the plastome (Chapter 4). We do not yet know enough about the structure-activity relationships of this protein to be able to predict exactly which amino acids to change, or indeed whether it will be possible to do this without affecting the carboxylation activity.

Nitrogen fixation

The development of new nitrogen-fixing plants is probably the genetic engineer's equivalent of the Philosopher's Stone, at least in terms of its intrinsic interest, if not in relation to its scientific validity. A great deal is known about the *Rhizobium*–legume system (Chapter 7). Many plant and bacterial genes are involved in the processes of recognition, infection, nodule development and function. A special subcellular structure and organization is required and the biochemical activity is integrated with that of the host plant. We probably need to know more about the molecular processes involved in recognition and interaction between plants and bacteria before we can assess whether it will be possible to arrange nitrogen-fixing associations with other crop plants. Although it presents a formidable challenge, the fact that we can begin to break the process down into a series of defined steps should at least provide the means for a rational approach to the problem.

Resistance to pests and diseases

The factors governing the resistance of plants to pests and diseases are complex (Callow, 1982) and we know very little about the genes involved in these processes. Plants show a general resistance to pathogens, which is related to the properties of surface waxes, hairs, the cuticle and suberized layers. In addition, they may contain compounds such as alkaloids and phenolics in the cytoplasm or vacuoles, which are inhibitory to fungal growth, and inhibitors of fungal enzymes in their walls. Plants may react to infection by local cell death, or by the production of physical barriers, enzymes such as chitinases which attack fungi, or low molecular weight compounds called phytoalexins, which inhibit fungal growth. Sometimes, specified fungal cell wall products (elicitors) can induce phytoalexin production. Genetic analysis shows that there is frequently a one-to-one relationship between virulence-avirulence genes in pathogens and resistance-susceptibility genes in the host. This is interpreted to mean that there are specific recognition processes between cell surfaces. Resistance and susceptibility are not confined to cell surfaces, however. Susceptibility of maize to the T-toxin of the fungus *Helminthosporium maydis* race T is associated with the mitochondrial genome (Chapter 5). Some progress has been made in studying the regulation of enzymes which may be involved in resistance to fungi (Chapter 6). Broadly speaking, our present knowledge of the biochemistry and molecular biology of resistance is inadequate for genetic engineering.

10.4 Control of expression of transferred genes

The control of gene expression plays an important part in plant development and it is obvious that consideration has to be given to the regulation of genes introduced artificially. The addition of the appropriate DNA signals for RNA polymerase binding, initiation of transcription, polyadenylation etc. seems relatively straightforward since the rules for these are becoming clear. However, the location of the genes and temporal and spatial aspects of their expression may also be important. Genes inserted near to heterochromatin may be inactivated by 'position effects' or by base methylation. In one series of experiments for example, a transferred gene was not expressed unless DNA methylation in the plant cells was inhibited (Hepburn *et al.*, 1983). It is also necessary to arrange for the gene product to appear in the correct place. In the case of atrazine resistance, for example, this means transferring the gene to the plastome (by a method not yet developed) and placing it under the same DNA control sequences as other photogenes or, alternatively, placing it in the nucleus. This seems rather a tall order at the present time, since it will be necessary to add a specifically-tailored DNA sequence coding for a transit peptide (Chapter 4), to alter the transcription and translation signals to the eukaryotic type, and to place the gene under the same DNA control sequences as the small subunit of rubisco or the light-harvesting protein (Chapter 6).

In some situations, however, it may be acceptable to have nuclear encoded genes which are involved in functions such as detoxification or resistance, switched on all the time. Sometimes, however, it is obviously desirable to be able to ensure the expression of a gene at a particular developmental stage such as flowering, grain-filling or in response to environmental signals such as heat-shock or photoperiod, or in response to plant growth substances. We do not yet know what the control sequences are for flowering genes, grain-filling genes etc., but this may become clear when more genomic DNA sequences are available for study.

10.5　Conclusions

A number of vector systems are now becoming available for the transfer of foreign genes to plants. In some cases, genes can be integrated into the chromosomes of the recipient and are expressed and stably inherited. Although it is not yet possible to transfer genes to all important groups of crop plants, the investigation of existing genetic systems may lead to the development of suitable vectors. The main problems faced by the genetic engineer are concerned with the identification and regulation of suitable genes for transfer. At the present time, there is an urgent need for fundamental studies on the biochemistry and molecular biology of plant processes in order to identify important genes. There is an equally pressing need to study the regulation of gene expression during development, in order to know how to turn genes on in the right place at the right time.

REFERENCES

Chapter 1

Further Reading

Grossman, L. and Moldave, K. (eds.) (1980) *Methods in Enzymology* **65**: *Nucleic Acids, Part 1*. Academic Press, New York etc.

Old, R. W. and Primrose, S. B. (1981) *Principles of Gene Manipulation: An Introduction to Genetic Engineering*, 2nd edn. Blackwell Scientific Publications, Oxford.

Rodriguez, R. L. and Tait, R. C. (1983) *Recombinant DNA Techniques*: *An Introduction*. Addison-Wesley Publishing Company, Reading, Massachusetts.

Walker, J. and Gaastra, W. (eds.) (1983) *Techniques in Molecular Biology*. Croom Helm, London.

Wu, R. (ed.) (1980) *Methods in Enzymology* **68**: *Recombinant DNA*. Academic Press, New York etc.

Wu, R., Grossman, L. and Moldave, K. (eds.) (1983) *Methods in Enzymology* **100**: *Recombinant DNA, Part B*. Academic Press, New York etc.

Wu, R., Grossman, L. and Moldave, K. (eds.) (1983) *Methods in Enzymology* **101**: *Recombinant DNA, Part C*. Academic Press, New York etc.

Chapter 2

Cited References

Appels, R. (1983) Chromosome structure in cereals: the analysis of regions containing repeated sequence DNA and its application to the detection of alien chromosomes introduced into wheat. In *Genetic Engineering of Plants: An Agricultural Perspective* (eds. T. Kosuge, C. P. Meredith and A. Hollaender). Plenum Press, New York etc., pp. 229–256.

Bazetroux, J., Jouanin, L. and Huguet, T. (1978) Characterisation of inverted repeated sequences in wheat nuclear DNA. *Nucl. Acid Res.* **5**, 751–769.

Bennet, M. D. and Smith, J. B. (1976) Nuclear DNA amounts in Angiosperms. *Phil. Trans. Roy. Soc. London Ser. B* **274**, 227–274.

Bryant, J. A. (1982) DNA replication and the cell cycle. In *Nucleic Acids and Proteins in Plants II. Structure, Biochemistry and Physiology of Nucleic Acids* (eds. B. Parthier and D. Boulter). Springer-Verlag Berlin etc., pp. 75–110.

Cullis, C. A. (1983) Environmentally induced DNA changes in plants. *CRC Critical Reviews in Plant Science* **1**, 117–131.

De Lange, R. J., Fambrough, D. M., Smith, E. L. and Bonner, J. (1969) Calf and pea histone 4. II Complete amino acid sequence of pea seedling histone 4; comparison with the homologous calf thymus histone. *J. Biol. Chem.* **224**, 5669–5679.

De Lange, R. J., Hooper, J. and Smith, E. L. (1973) Histone 3. III Sequence studies on the cyanogen bromide peptides; complete amino acid sequence of calf thymus histone 3. *J. Biol. Chem.* **248**, 3261–3274.

Döring, H. P., Tillman, E. and Starlinger, P. (1984) DNA sequence of the maize transposable element *Dissociator. Nature* **307**, 127–130.

Flavell, R. B. (1982) Chromosomal DNA sequences and their organisation. In *Nucleic Acids and Proteins in Plants II. Structure, Biochemistry and Physiology of Nucleic Acids* (eds. B. Pathier and D. Boulter). Springer-Verlag, Berlin etc., pp. 46–74.

Flavell, R. B. (1983) Chromosomal variation at the molecular level in crop plants. In *Genetic Engineering: Applications to Agriculture* (ed. L. D. Owens). Rowman and Allanheld, Ottawa, pp. 15–25.

Gerlach, W. L. and Dyer, T. A. (1980) Sequence organisation of the repeating units in the nucleus of wheat which contain 55 rRNA genes. *Nucl. Acid Res.* **8**, 4851–4865.

Goldberg, R. B., Hoschek, G. and Kamalay, J. C. (1978) Sequence complexity of nuclear and polysomal RNA in leaves of the tobacco plant. *Cell* **14**, 123–131.

Grierson, D. (1982) RNA processing and other post-transcriptional modifications. In *Nucleic Acids and Proteins in Plant II. Structure, Biochemistry and Physiology of Nucleic Acids* (eds. B. Parthier and D. Boulter). Springer-Verlag, Berlin etc., pp. 192–223.

Hemleben, V., Grierson, D. and Dertmann, H. (1977). The use of equilibrium centrifugation in actinomycin-caesium chloride for the purification of ribosomal DNA. *Plant Sci. Lett.* **9**, 129–135.

Ingle, J., Pearson, G. C. and Sinclair, J. (1973) Species distribution and properties of nuclear satellite DNA in higher plants. *Nature New Biol.* **242**, 193–197.

McClintock, B. (1946) Maize genetics. *Carnegie Inst. Wash. Yb.* **45**, 176–186.

McClintock, B. (1951) Chromosome organisation and genic expression. *Cold Spring Harb. Symp. Quant. Biol.* **16**, 13–47.

Murray, M. G., Cuellar, R. E. and Thompson, W. F. (1978) DNA sequence organisation in the pea genome. *Biochemistry* **17**, 5781–5790.

Shepherd, N. S., Schwarz-Sommer, Z., Vel Spalve, J. B., Gupta, M., Wienand, U. and Saedler, H. (1984). Similarity of the *Cin* I repetitive family of *Zea mays* to eukaryotic transposable elements. *Nature* **307**, 185–187.

Van't Hof, J. and Bjerknes, C. A. (1979) Chromosomal DNA replication in higher plants. *Bioscience* **29**, 18–22.

Chapter 3

Cited References

Cashmore, A. R. (1983) Nuclear genes encoding the small subunit of ribulose-1,5-bisphosphate carboxylase. In *Genetic Engineering of Plants: An Agricultural Perspective* (eds. T. Kosuge, C. P. Meredith and A. Hollaender). Plenum Press, New York, pp. 29–38.

Grierson, D. (1977) The nucleus and the organisation and transcription of nuclear DNA. In *Molecular Biology of Plant Cells* (ed. H. Smith). Blackwell Scientific Publications, Oxford, pp. 213–255.

Grierson, D. (1982) RNA processing and other post-transcriptional modifications. In *Nucleic Acids and Proteins in Plants. II Structure, Biochemistry and Physiology of Nucleic Acids* (eds. B. Pathier and D. Boulter). Springer-Verlag, Berlin etc., pp. 192–223.

Guerrier-Takada, G., Gardiner, K., March, T., Pace, N. and Altman, S. (1983) The RNA moiety of ribonuclease P is the catalytic subunit of the enzyme. *Cell* **35**, 849–857.

Filipowicz, W. and Gros, H. J. (1984) RNA ligation in eukaryotes. *Trends Biochem. Sci.* **9**, 68–71.

Langridge, P. and Feix, G. (1983) A zein gene of maize is transcribed from two widely separated promoter regions. *Cell* **34**, 1015–1022.

Larkins, B. A. (1983) Genetic engineering of seed storage proteins. In *Genetic Engineering of Plants: An Agricultural Perspective* (eds. T. Kosuge, C. P. Meredith and A. Hollaender) Plenum Press, New York, pp. 93–118.

Messing, J., Geraghty, D., Heidecker, G., Hu, N-T., Kridl, J. and Rubenstein, I. (1983) Plant gene structure. In *Genetic Engineering of Plants: An Agricultural Perspective* (eds. T. Kosuge, C. P. Meredith and A. Hollaender). Plenum Press, New York, pp. 211–227.

Sänger, H. L. (1982) Biology, structure, function and possible origin of viroids. In *Nucleic Acids and Proteins in Plants II Structure, Biochemistry and Physiology of Nucleic Acids* (eds. B. Parthier and D. Boulter). Springer-Verlag, Berlin etc., pp. 368–454.

Shah, D. P., Hightower, R. C. and Meagher, R. B. (1982) Complete nucleotide sequence of a soybean actin gene. *Proc. Natl. Acad. Sci. USA* **79**, 1022–1026.

Chapter 4

Cited References

Blair, G. E. and Ellis, R. J. (1973) Protein synthesis in chloroplasts. 1. Light-driven synthesis of the large subunit of Fraction I protein by isolated pea chloroplasts. *Biochim. Biophys. Acta* **319**, 223–234.

Bogorad, L., Gubbins, E. J., Krebbers, E. T., Larrinua, I. M., Muskavitch, K. M. T., Rodermel, S. R. and Steinmetz, A. (1983). The organization and expression of maize plastid genes. In *Genetic Engineering: Applications to Agriculture* (ed. L. D. Owens). Rowman and Allanheld, Ottawa, pp. 35–53.

Bohnert, H. J., Crouse, E. J., and Schmitt, J. M. (1982). Organisation and expression of plastid genomes. In *Nucleic Acids and Proteins in Plants II. Structure, Biochemistry and Physiology of Nucleic Acids* (eds. B. Parthier and D. Boulter). Springer-Verlag, Berlin etc., pp. 475–530.

Bottomley, W. and Bohnert, H. J. (1982) The biosynthesis of chloroplast proteins. In *Nucleic Acids and Proteins in Plants II. Structure, Biochemistry and Physiology of Nucleic Acids* (eds. B. Parthier and D. Boulter). Springer-Verlag, Berlin etc., pp. 531–596.

Bottomley, W., Spencer, D. and Whitfeld, P. R. (1974) Protein synthesis in isolated spinach chloroplasts: Comparison of light-driven and ATP-driven synthesis. *Arch. Biochem. Biophys.* **164**, 120–124.

Criddle, R. S., Dau, B., Kleinkopf, G. E. and Huffaker, R. C. (1970) Differential synthesis of ribulose diphosphate carboxylase subunits. *Biochim. Biophys. Res. Commun.* **41**, 621–627.

Dobberstein, B., Blobel, G. and Chua, N-H. (1977) *In vitro* synthesis and processing of a putative precursor for the small subunit of ribulose-1,5-bisphosphate carboxylase of *Chlamydomonas reinhardtii. Proc. Natl. Acad. Sci. USA* **74**, 1082–1085.

Doherty, A. and Gray, J. C. (1980) Synthesis of a dicyclohexylcarbodiimide-binding proteolipid by isolated pea chloroplasts. *Eur. J. Biochem.* **108**, 131–136.

Dyer, T. A. (1982) RNA sequences. In *Nucleic Acids and Proteins in Plants II. Structure, Biochemistry and Physiology of Nucleic Acids* (eds. B. Parthier and D. Boulter), Springer-Verlag, Berlin etc., pp. 171–191.

Edwards, K. and Kössel, H. (1981) The rRNA operon from *Zea mays* chloroplasts: nucleotide sequences of 23S rDNA and its homology with *E. coli* 23S rDNA. *Nucleic Acids Res.* **9**, 2853–2869.

Ellis, R. J. (1981) Chloroplast proteins: synthesis, transport and assembly. *Ann. Rev. Plant Physiol.* **32**, 111–137.

Ellis, R. J. (1983) Mobile genes of chloroplasts and the promiscuity of DNA. *Nature* **304**, 308–309.

Grierson, D. (1982) RNA processing and other post-transcriptional modifications. In *Nucleic Acids and Proteins in Plants II. Structure, Biochemistry and Physiology of Nucleic Acids* (eds. B. Parthier and D. Boulter). Springer-Verlag, Berlin etc., pp. 192–223.

Grossman, A. R., Bartlett, S. G., Schmidt, G. W., Mullet, J. E. and Chua, N-H. (1982) Optimal conditions for the post-translational uptake of proteins by isolated chloroplasts. *In vitro* synthesis and transport of plastocyanin, ferridoxin-NADP$^+$ oxidoreductase and fructose 1,6-bisphosphatase. *J. Biol. Chem.* **257**, 1558–1563.

Heinhorst, S., and Shively, J. M. (1983) Encoding of both subunits of ribulose 1,5-bisphosphate carboxylase by organelle genome of *Cyanophora paradoxa. Nature* **304**, 373–374.

Highfield, P. E. and Ellis, R. J. (1978) Synthesis and transport of the small subunit of ribulose bisphosphate carboxylase. *Nature* **271**, 420–424.

Kawashima, N. and Wildman, S. G. (1972) Studies on fraction 1 protein. IV Mode of inheritance of primary structure in relation to whether chloroplast or nuclear DNA contains the code for a chloroplast protein. *Biochim. Biophys. Acta* **262**, 42–49.

Kolodner, R. and Tewari, K. K. (1975) Chloroplast DNA from higher plants replicates by both the Cairns and the rolling circle mechanism. *Nature* **256**, 708–711.

Lyttleton, J. W. (1962) Isolation of ribosomes from spinach chloroplasts. *Exp. Cell Res.* **26**, 312–317.

McIntosh, L., Poulsen, C. and Bogorad, L. (1980) Chloroplast gene sequence for the large subunit of ribulose bisphosphate carboxylase of maize. *Nature* **288**, 556–560.

Nelson, N., Nelson, H. and Schatz, G. (1980) Biosynthesis and assembly of the proton-translocating adenosine triphosphatase complex from chloroplasts. *Proc. Natl. Acad. Sci USA* **77**, 1361–1364.

Reid, R. A. and Leech, R. M. (1980) *Biochemistry and Structure of Cell Organelles.* Blackie, Glasgow and London.

Ris, H. and Plant, W. (1962) The ultrastructure of DNA-containing areas in the chloroplast of *Chlamydomonas. J. Cell Biol.* **13**, 383–391.

Schwartz, Z. and Kössel, H. (1980) The primary structure of 16S rDNA from *Zea mays* chloroplasts is homologous with *E. coli* 16S rRNA. *Nature* **283**, 739–742.

Scott, N. S. and Smillie, R. M. (1967) Evidence for the direction of chloroplast ribosomal RNA synthesis by chloroplast DNA. *Biochem. Biophys. Res. Comm.* **28**, 598–603.

Scott, N. S. and Possingham, J. V. (1982) Leaf development. In *The Molecular Biology of Plant Development*, (eds. H. Smith and D. Grierson). Blackwell Scientific Publications, Oxford, pp. 223–255.

Steinback, K. E., McIntosh, L., Bogorad, L. and Arntzen, C. J. (1981) Identification of the triazene receptor protein as a chloroplast gene product. *Proc. Nat. Acad. Sci. USA* **78**, 7463–7467.

Takaiwa, F. and Sugiura, M. (1982) The complete nucleotide sequence of a 23S rRNA gene from tobacco chloroplasts. *Eur. J. Biochem.* **124**, 13–19.

Tohdoh, N. and Sugiura, M. (1982) The complete nucleotide sequence of 16S ribosomal RNA gene from tobacco chloroplasts. *Gene* **17**, 213–218.

Wollgiehn, R. (1982) *RNA polymerase and regulation of transcription.* In *Nucleic Acids and Proteins in Plants II. Structure, Biochemistry and Physiology of Nucleic Acids.* (eds. B. Parthier and D. Boulter). Springer-Verlag, Berlin etc., pp. 125–170.

Chapter 5

Cited References

Carignani, G., Groudinsky, O., Frezza, D., Schiavon, E., Bergantino, E. and Slonimski, P. P. (1983) An mRNA maturase is encoded by the first intron of the mitochondrial gene for the subunit 1 of cytochrome oxidase in *S. cerevisiae*. *Cell* **35**, 733–742.

Chao, S., Sederoff, R. R. and Levings, C. S. (1983) Partial sequence analysis of the 5S to 18S rRNA gene region of the maize mitochondrial genome. *Plant Physiol.* **71**, 190–193.

Ellis, R. J. (1982) Promiscuous DNA-chloroplast genes inside plant mitochondria. *Nature* **299**, 678–679.

Forde, B. G. and Leaver, C. J. (1979) Nuclear and cytoplasmic genes controlling synthesis of variant mitochondrial polypeptides in male-sterile maize. *Proc. Nat. Acad. Sci. USA* **77**, 418–422.

Fox, T. D. and Leaver, C. J. (1981) Maize mitochondrial cytochrome oxidase II subunit has an intervening sequence and does not contain TGA codons. *Cell* **26**, 315–323.

Hack, E. and Leaver, C. J. (1983) The α-subunit of the maize F_1-ATPase is synthesised in the mitochondrion. *EMBO J.* **2**, 1783–1789.

Jukes, T. H. (1983) Mitochondrial codes and evolution. *Nature* **301**, 19–20.

Kück, U., Stahl, U. and Esser, K. (1981) Plasmid-like DNA is part of mitochondrial DNA in *Podospera anserina*. *Current Genet.* **3**, 151–156.

Leaver, C. J. and Gray, M. W. (1982) Mitochondrial genome organisation and expression in higher plants. *Ann. Rev. Plant Physiol.* **33**, 373–402.

Leaver, C. J., Hack, E., Dawson, A. J., Isaac, P. G. and Jones, U. P. (1983) Mitochondrial genes and their expression in higher plants. In *Nucleo-mitochondrial Interactions* (eds. R. J. Schweyen, K. Wolf and K. Kaudewitz). Walter de Gruyter, Berlin etc., pp. 269–283.

Palmer, J. D. and Shields, C. R. (1984) Tripartite structure of the *Brassica campestris* mitochondrial genome. *Nature* **317**, 437–440.

Schatz, G. and Butow, R. A. (1983) How are proteins imported into mitochondria. *Cell* **32**, 316–318.

Stern, D. D. and Lonsdale, D. M. (1982) Mitochondrial and chloroplast genomes of maize have a 12-kilobase DNA sequence in common. *Nature* **299**, 699–702.

Chapter 6

Cited References

Apel, K. (1979) Phytochrome-induced appearance of mRNA activity for the apoprotein of the light harvesting chlorophyll *a/b* protein of barley (*Hordeum vulgare*) *Eur. J. Biochem.* **97**, 183–188.

Bennett, J. (1983) Regulation of photosynthesis by reversible phosphorylation of the light-harvesting protein. *Biochem. J.* **212**, 1–13.

Bogorad, L., Gubbins, E. J., Krebbers, E. T., Larrinua, I. M. Muskavitch, K. M. T., Rodermel, S. R. and Steinmetz, A. (1983) The organisation and expression of maize plastid genes. In *Genetic Engineering: Applications to Agriculture* (ed. L. D. Owens). Rowman and Allanheld, Ottawa, pp. 35–53.

Boston, R. S., Miller, T. J., Mertz, J. E. and Burgess, R. R. (1982) *In vitro* synthesis and processing of wheat α-amylase. Translation of gibberellic acid-induced wheat aleurone layer RNA by wheat germ and *Xenopous laevis* oocyte systems. *Plant Physiol.* **69**, 150–154.

Edelman, M. and Reisfeld, A. (1980) Synthesis processing and functional probing of P-32000, the major protein translated within the chloroplast. In *Genome Organisation and Expression in Plants* (ed. C. J. Leaver). Plenum Press, New York etc., pp. 353–362.

Gallagher, T. E. and Ellis, R. J. (1982) Light-stimulated transcription of genes for two chloroplast polypeptides in isolated pea leaf nuclei. *EMBO Journal* **1**, 1493–1498.

Grierson, D. (1984) Nucleic acid and protein synthesis during fruit ripening and senescence. In *Cell Aging and Cell Death* (eds. I. Davies and D. C. Sigee). Cambridge University Press (in press).

Grierson, D., Tucker, G. A. and Robertson, N. G. (1981) The molecular biology of ripening. In *Recent Advances in the Biochemistry of Fruit and Vegetables* (eds. J. Friend and M. J. C. Rhodes). Academic Press, London etc., pp. 149–160.

Hahlbrock, K., Boudet, A. M., Chappell, J., Kreuzaler, F., Khun, D. K. and Ragg, H. (1983) Differential induction of mRNAs by light and elicitor in cultured plant cells. In *Structure and Function of Plant Genomes* (eds. O. Cifferi and L. Dure). Plenum Press, New York etc., pp. 15–23.

Higgins, T. J. V., Zwar, J. A. and Jacobsen, J. V. (1976) Gibberellic acid enhances the level of translatable mRNA for α-amylase in barley aleurone layers. *Nature* **260**, 166–169.

Jones, R. L. and Jacobsen, J. U. (1980) The role of the endoplasmic reticulum in the synthesis and transport of α-amylase in barley aleurone layers. *Planta* **156**, 421–432.

Muthukrishnan, S. and Chandra, G. R. (1983) Expression of α-amylase genes in barley aleurone cells. In *Genetic Engineering: Applications to Agriculture* (ed. L. D. Owens). Rowman and Allanheld, Ottawa, pp. 151–159.

Rosner, A., Jakob, K. M., Gressel, J. and Sagher, D. (1975). The early synthesis and possible function of a 0.5×10^6 M_r RNA after transfer of dark grown *Spirodela* plants to light. *Biochem. Biophys. Res. Commun.* **67**, 383–391.

Schröder, J., Kreuzaler, F., Schäfer, E. and Hahlbrock, K. (1979) Concomitant induction of phenylalanine ammonia-lyase and flavone synthase mRNAs in irradiated plant cells *J. Biol. Chem.* **245**, 57–65.

Further reading

Leaver, C. J. (ed.) (1980) *Genome Organisation and Expression in Plants.* Plenum Press, New York etc.

Cifferi, O. and Dure, L. (eds.) (1983) *Structure and Function of Plant Genomes.* Plenum Press, New York etc.

Chapter 7

Cited References

Baulcombe, H. and Verma, D. P. S. (1978) Preparation of a complementary DNA for leghemoglobin and direct demonstration that leghemoglobin is encoded by the soybean genome. *Nucleic. Acids Res.* **5**, 4141–4153.

Bergmann, H., Preddie, E. and Verma, D. P. S. (1983) Nodulin-35: a subunit of specific uricase (uricase II) induced and localized in the uninfected cells of soybean nodules. *EMBO J.* **2**, 2333–2339.

Brisson, N. and Verma, D. P. S. (1982) Soybean leghemoglobin gene family: normal, pseudo and truncated genes. *Proc. Natl. Acad. Sci. USA* **79**, 4055–4059.

Cullimore, J. V., Lara, M., Lea, P. J. and Miflin, B. J. (1983) Purification and properties of two forms of glutamine synthetase from the plant fraction of *Phaseolus* root nodules. *Planta* **157**, 245–253.

Downie, J. A., Hombrecher. G., MA, Q-S., Knight, C. D., Wells, B. and Johnston, A. B. W. (1983*a*) Cloned nodulation genes of *Rhizobium leguminosarum* determine host-range specificity. *Molec. Gen. Genet.* **190**, 359–365.

Downie, J. A., Ma, Q-S., Knight, C. D., Hombrecher, G. and Johnston, A. B. W. (1983*b*)

Cloning of the symbiotic region of *Rhizobium leguminosarum*: the nodulation genes are between the nitrogenase genes and a *nifA*-like gene. *EMBO J.*, **2**, 947–952.

Fuller, F., Kunster, P. W., Nguyen, T. and Verma, D. P. S. (1983) Soybean nodulin genes: analysis of cDNA clones reveals several major tissue-specific sequences in nitrogen-fixing root nodules. *Proc. Natl. Acad. Sci. USA* **80**, 2594–2598.

Fuller, F. and Verma, D. P. S. (1984) Appearance and accumulation of nodulin mRNAs and their relationship to the effectiveness of root nodules. *Plant Molec. Biol.* **3**, 21–28.

Hirsch, P. R., Van Montagu, M., Johnston, A. B.W., Brewin, N. J. and Schell, J. (1980) Physical identification of bacteriocinogenic, nodulation and other plasmids in strains of *Rhizobium leguminosarum*. *J. Gen. Microbiol.* **120**, 403–412.

Hyldig-Nielsen, J. J., Jensen, E. O., Paludan, K., Wiborg, O., Garrett, R., Jorgensen, P. and Marcker, K. A. (1982) The primary structures of two leghemoglobin genes from soybean. *Nucleic Acids Res.* **10**, 689–701.

Johnston, A. B.W., Beynon, J. L., Buchanan-Wollaston, A. V., Setchell, S. M., Hirsch, P. R. and Beringer, J. E. (1978) High frequency transfer of nodulating ability between strains and species of *Rhizobium*. *Nature* **276**, 634–636.

Meijer, E. G. M. and Broughton, W. J. (1982) Biology of legume-*Rhizobium* interactions in nodule formation. In *Molecular Biology of Plant Tumours* (eds. G. Khal and J. Schell). Academic Press, New York, pp. 107–129.

Nuti, M. P., Ledeboer, A. M., Lepidi, A. A. and Schilperoort, R. A. (1977) Large plasmid in different *Rhizobium* species. *J. Gen. Microbiol.* **100**, 241–248.

Nuti, M. P., Lepidi, A. A., Prakash, R. K., Hooykaas, P. J. J. and Schilperoort, R. A. (1982) The plasmids of *Rhizobium* and symbiotic nitrogen fixation. In *Molecular Biology of Plant Tumours* (eds. G. Khan and J. Schell). Academic Press, New York, pp. 561–588.

Vance, C. P. (1983) *Rhizobium* infection and nodulation: a beneficial plant disease? *Ann. Rev. Microbiol.* **37**, 399–424.

Verma, D. P. S., Nash, D. T. and Schulman, H. M. (1974) Isolation and *in vitro* translation of soybean leghemoglobin mRNA. *Nature* **251**, 74–77.

Verma, D. P. S., Haughland, R., Brisson, N., Legocki, R. P. and Lacroix, L. (1981) Regulation of the expression of leghaemoglobin genes in effective and ineffective root nodules of soybean. *Biochim. Biophys. Acta* **653**, 98–107.

Vincent, J. M. (1980) Factors controlling the legume-*Rhizobium* symbiosis. In *Nitrogen Fixation II* (eds. W. E. Newton and W. H. Orme-Johnson). University Park Press, Baltimore pp. 103–129.

Wiborg, O., Hyldig-Nielsen, J. J., Jensen, E. O., Paludan, K. and Marcker, K. A. (1983) The structure of an unusual leghemoglobin gene from soybean. *EMBO J.* **2**, 449–452.

Further Reading

Stewart, W. P. D. and Gallon, J. R. (eds.) (1980) *Nitrogen Fixation*. Academic Press, London.

Verma, D. P. S. and Long, S. (1983) The molecular biology of *Rhizobium*-legume symbiosis. *Int. Rev. Cytol Supp.* **14**, (ed. K. Jeon). Academic Press, New York, pp. 211–245.

Chapter 8

Cited References

Barker, R. F., Idler, K. B., Thompson, D. V. and Kemp, J. D. (1983) Nucleotide sequence of the T-DNA region from *Agrobacterium tumefaciens* octopine Ti plasmid pTi 15955. *Plant Molec. Biol.* **2**, 335–350.

Bevan, M. W. and Chilton, M-D. (1982) T-DNA of the *Agrobacterium* Ti and Ri plasmids. *Ann. Rev. Genet.* **16**, 357–384.

Braun, A. C. (1982) A history of the crown gall problem. In *Molecular Biology of Plant Tumours* (eds. G. Kahl and J. Schell). Academic Press, New York, pp. 155–210.

Caplan, A., Herrera-Estrella, L., Inze, D., Van Haute, E., Van Montagu, M., Schell, J. and Zambryski, P. (1983) Introduction of genetic material into plant cells. *Science* **222**, 815–821.

Chilton, M-D., Drummond, M. H., Merlo, D. J., Sciaky, D., Montoya, A. L., Gordon, M. P. and Nester, E. W. (1977) Stable incorporation of plasmid DNA into higher plant cells: the molecular basis of crown gall tumourigensis. *Cell* **11**, 263–271.

Depicker., A., De Wilde, M., De Vos, G., De Vos, R., Van Montagu, M. and Schell, J. (1980) Molecular cloning of the nopaline Ti Plasmid pTi C58 and its use for restriction endonuclease mapping. *Plasmid* **3**, 193–211.

De Vos, G., De Beuckeleer, M., Van Montagu, M. and Schell, J. (1981) Addendum. Restriction endonuclease mapping of the octopine tumour-inducing plasmid pTi Ach 5 of *Agrobacterium tumefaciens*. *Plasmid* **6**, 249–253.

Hamilton, R. H. and Fall, M. Z. (1971) The loss of tumour-initiating ability in *Agrobacterium tumefaciens* by incubation at high temperature. *Experientia* **27**, 229–230.

Kerr, A. (1969) Transfer of virulence between strains of *Agrobacterium*. *Nature* **223**, 1175–1176.

Murai, N. and Kemp, J. D. (1982) Octopine synthase mRNA isolated from sunflower crown gall callus is homologous to the Ti plasmid of *Agrobacterium tumefaciens*. *Proc. Natl. Acad. Sci. USA* **79**, 86–90.

Nester, E. W. and Kosuge, T. (1981) Plasmids specifying plant hyperplasias. *Ann. Rev. Microbiol.* **35**, 531–565.

Petit, A., Delhaye, S., Tempe, J. and Model, G. (1970) Recherches sur les guanidines des tissus de crown gall. Mise en évidence d'une relation biochimique spécifique entre les souches d'*Agrobacterium tumefaciens* et les tumeurs qu'elles induisent. *Physiol. Veg.* **8**, 205–213.

Schell, J., Van Montagu, M., De Beuckeleer, M., De Block, M., Depicker, A., De Wilde, M., Engler, G., Genetello, C., Hernalsteens, J. P., Holsters, M., Seurinck, J., Silva, B., Van Vliet, F. and Villarroel, R. (1979) Interactions and DNA transfer between *Agrobacterium tumefaciens*, the Ti plasmid and the plant host. *Proc. Roy. Soc. Lond. Ser. B* **204**, 251–266.

Van Larebeke, N., Gentello, C., Schell, J., Schilperoort, R. A., Hermans, A. K., Hernalsteens, J. P. and Van Montagu, M. (1975) Acquisition of tumour-inducing ability by non-oncogenic agrobacteria as a result of plasmid transfer. *Nature* **255**, 742–743.

Watson, B., Currier, T. C., Gordon, M. D., Chilton, M-D. and Nester, E. W. (1975) Plasmid required for virulence of *Agrobacterium tumefaciens*. *J. Bacteriol.* **123**, 255–264.

Yadav, N. S., Vanderleyden, J., Bennett, D. R., Barnes, W. M. and Chilton, M-D. (1982) Short direct repeats flank the T-DNA on the nopaline Ti plasmid. *Proc. Natl. Acad. Sci. USA* **79**, 6322–6326.

Zaenen, I., Van Larebeke, N., Teuchy. N., Van Montagu, M. and Schell, J. (1974) Supercoiled circular DNA in crown gall inducing *Agrobacterium* strains. *J. Molec. Biol.* **86**, 109–127.

Zambryski, P., Depicker, A., Kruger, K. and Goodman, H. (1982) Tumour induction by *Agrobacterium tumefaciens*: analysis of the boundaries of T-DNA. *J. Molec. Appl. Genet.* **1**, 361–370.

Further Reading

Kahl, G. and Schell, J. (eds.) *Molecular Biology of Plant Tumours*. Academic Press, New York, 1982.

Zambryski, P., Goodman, H., Van Montagu, M. and Schell, J. (1983) *Agrobacterium* tumour induction. In *Mobile Genetic Elements* (ed. J. Shapiro). Academic Press, New York, pp. 505–535.

Chapter 9

Ahlquist, P., Dasgupta, R. and Kaesberg, P. (1984) Nucleotide sequence of the brome mosaic virus genome and its implications for viral replication. *J. Molec. Biol.* **172**, 369 – 383.

Davies, J. W. and Hull, R. (1982) Genome expression of plant positive-strand viruses. *J. Gen. Virol.* **61**, 1–14.

Franck, A., Guilley, H., Jonard, G., Richards, K. and Hirth, L. (1980) Nucleotide sequence of cauliflower mosaic virus DNA. *Cell* **21**, 285–294.

Goelet, P., Lomonossoff, G. P., Butler, P. J. G., Akam, M. E., Gait, M. J. and Karan, J. (1982) Nucleotide sequence of tobacco mosaic virus RNA. *Proc. Natl. Acad. Sci. USA* **79**, 5818–5822.

Goodman, R. M. (1981) Geminiviruses. *J. Gen. Virol.* **54**, 9– .

Guilley, H., Jonard, G., Kukla, B. and Richards, K. E. (1979) Sequence of 1000 nucleotides at the 3′ end of tobacco mosaic virus RNA. *Nucleic Acids Res.* **6**, 1287–1308.

Guilley, H., Richards, K. E. and Jonard, G. (1983) Observations concerning the discontinuous DNAs of cauliflower mosaic virus. *EMBO J.* **2**, 277–282.

Hohn. T., Richards, K. E. and Lebeurier, G. (1982) Cauliflower mosaic virus on its way to becoming a useful plant vector. *Curr. Top. Microbiol. Immunol.* **96**, 193–236.

Hull, R. and Covey, S. N. (1983) Replication of cauliflower mosaic virus DNA. *Sci. Prog. Oxf.* **68**, 403–422.

Hull, R. and Davies, J. W. (1983) Genetic engineering with plant viruses, and their potential as vectors. *Adv. Virus Res.* **28**, 1–33.

Kiefer, M. C., Owens, R. A. and Diener, T. O. (1983) Structural similarities between viroids and transposable genetic elements. *Proc. Natl. Acad. Sci. USA* **80**, 6234–6238.

Kozak, M. (1983) Comparison of initiation of protein synthesis in procaryotes, eucaryotes and organelles. *Microbiol. Rev.* **47**, 1–45.

Lomonossoff, G. P. and Shanks, M. (1983) The nucleotide sequence of cowpea mosaic virus B RNA. *EMBO J.* **2**, 2253–2258.

Matthews, R. E. F. (1982) Classification and nomenclature of viruses. *Intervirology* **17**, nos. 1–3.

Olszewski, N., Hagen, G. and Guilfoyle, T. J. (1982) A transcriptionally active minichromosome of cauliflower mosaic DNA isolated from infected turnip leaves. *Cell* **29**, 395–402.

Robertson, H. D., Howell, S. H., Zaitlin, M. and Malmberg, R. L. (1983) (eds.) *Plant Infectious Agents: Viruses, Viroids, Virusoids and Satellites.* Cold Spring Harbour, New York.

Sänger, H. L. (1982) Biology, Structure, function and possible origin of viroids. In *Encyclopedia of Plant Physiology, New Series* vol. 14B (eds. B. Parthier and D. Boulter). Springer-Verlag, Berlin etc., pp. 368–454.

Stanley, J. and Gay, M. R. (1983) Nucleotide sequence of cassava latent virus DNA. *Nature* **301**, 260–262.

Turner, D. S. and Covey, S. N. (1984) A putative primer for the replication of cauliflower mosaic virus by reverse transcription is virion-associated. *FEBS Lett.* **165**, 285–289.

Van Wezenbeek, P., Verver, J., Harmsen, J., Vos, P. and Van Kammen, A. (1983) Primary structure and gene organisation of the middle component of cowpea mosaic virus. *EMBO J.* **2**, 941–946.

Further Reading

Davies, J. W. (ed). *Molecular Plant Virology*, vols. 1–3. CRC Press, Florida, in press.

Chapter 10

Cited References

Beversdorf, W. D., Weiss-Lerman, J. and Erickson, L. R. (1980) Registration of triazine-resistant *Brassica campestris* germ plasm. *Crop Science* **20**, 289.

Callow, J. A. (1982) Molecular aspects of fungal infections. In *The Molecular Biology of Plant*

Development, (eds. H. Smith and D. Grierson). Blackwell Scientific Publications, Oxford, pp. 467–497.

Caplan, A., Herrera-Estrella, L., Inze, D., Van Haute, E., Van Montagu M., Shell, J. and Zambryski, P. (1983) Introduction of genetic material into plant cells. *Science* **222**, 815–821.

Daubert, S., Shepherd, R. J. and Gardner, R. C. (1983) Insertional mutagenesis of the cauliflower mosaic virus genome. *Gene* **25**, 201–208.

De Framond, A. J., Barton, K. A. and Chilton, M-D. (1983) Mini Ti: a new vector strategy for plant genetic engineering. *Biotechnology* **1**, 262–269.

Dixon, L. K., Koenig, I. and Hohn, T. (1983) Mutagenesis of cauliflower mosaic virus. *Gene* **25**, 189–199.

Hepburn, A. G., Clarke, L. E., Pearson, L. and White, J. (1983) The role of cytosine methylation in the control of nopaline synthase gene expression in a plant tumour. *J. Mol. Appl. Genet.* **2**, 315–329.

Hernalsteens, J. P., Van Vliet, F., De Beuckeleer, M., Depicker, A., Engler, G., Lemmers, M., Holsters, M., Van Montagu, M. and Schell, J. (1980) The *Agrobacterium tumefaciens* Ti plasmid as a host vector system for introducing foreign DNA into plant cells. *Nature* **287**, 654–656.

Herrera-Estrella, L., Depicker, A., Van Montagu, M. and Schell, J. (1983) Expression of chimaeric genes transferred into plant cells using a Ti plasmid-derived vector. *Nature* **303**, 209–213.

Howell, S. H., Walker, L. L. and Dudley, R. K. (1980) Cloned cauliflower mosaic virus DNA infects turnips (*Brassica rapa*). *Science* **208**, 1265–1267.

Hull, R. (1978) The possible use of plant viral DNAs in genetic manipulation in plants. *Trends Biochem. Sci.* **3**, 254–256.

Hull, R. and Davies, J. W. (1983) Genetic engineering with plant viruses, and their potential as vectors. *Adv Virus Res.* **28**, 1–33.

Murai, N., Sutton, D. W., Murray, M. G., Slightom, J. L., Merlo, D. J., Reichert, N. A., Sengupta-Gopalan, C., Stock, C. A., Barker, R. F., Kemp, D. J. and Hall T. C. (1983) Phaseolin gene from bean is expressed after transfer to sunflower via tumour-inducing plasmid vectors. *Science* **222**, 476–482.

Racaniello, V. R. and Baltimore, D. (1981) Cloned poliovirus complementary DNA is infectious in mammlian cells. *Science* **214**, 916–919.

Stanley, J. (1983) Infectivity of the cloned geminivirus genome requires sequences from both DNAs. *Nature* **305**, 643–645.

Taniguchi, T., Palmieri, M. and Weissmann, C. (1978) Qβ DNA-containing hybrid plasmids give rise to Qβ phage formation in the bacterial host. *Nature* **274**, 223–228.

Walden, R. M. and Howell, S. H. (1982) Intergenomic recombination events among pairs of defective cauliflower mosaic virus genomes. *J. Mol. App. Genet.* **1**, 447–456.

Further reading

Arber, W., Illmensee, K., Peacock, W. J. and Starlinger, P. (eds.) (1984) *Genetic Manipulation, Impact on Man and Society*. Cambridge University Press, London.

Kosuge, T., Meredith, C. P. and Hollaender, A. (eds.) (1983) *Genetic Engineering of Plants: An Agricultural Perspective*. Plenum Press, New York etc.

Owens, L. D. (ed.) (1983) *Genetic Engineering: Applications to Argiculture*. Rowman and Allanheld, Ottawa.

TERTIARY LEVEL BIOLOGY

Methods in Experimental Biology	Ralph
Visceral Muscle	Huddart and Hunt
Biological Membranes	Harrison and Lunt
Comparative Immunobiology	Manning and Turner
Biology of Nematodes	Croll and Matthews
Biology of Ageing	Lamb
An Introduction to Marine Science	Meadows and Campbell
An Introduction to Developmental Biology	Ede
Physiology of Parasites	Chappell
Neurosecretion	Maddrell and Nordmann
Biology of Communication	Lewis and Gower
Population Genetics	Gale
Biochemistry and Structure of Cell Organelles	Reid
Developmental Microbiology	Peberdy
Genetics of Microbes	Bainbridge
Biological Functions of Carbohydrates	Candy
Endocrinology	Goldsworthy, Robinson and Mordue
The Estuarine Ecosystem	McLusky
Animal Osmoregulation	Rankin and Davenport
Molecular Enzymology	Wharton and Eisenthal
Environmental Microbiology	Grant and Long
The Genetic Basis of Development	Stewart and Hunt

In the USA, these titles are distributed by John Wiley and Sons, New York.

170

Index